W9-CNA-752

Cybertraps for the Young

Cybertraps for the Young

Frederick S. Lane

NTI Upstream  Chicago

NTI Upstream
180 N. Michigan Ave., Ste. 700
Chicago, Illinois 60601
Visit our website at www.ntiupstream.com

NTI Upstream books may be purchased for educational, business, or sales promotional use. For more information about special discounts for bulk purchases or to book a live event, please contact NTI Upstream Special Sales at 1-312-423-5680.

Grateful acknowledgment is made to the National Institute on Drug Abuse for use of material adapted from "Monitoring the Future Study: Trends in Prevalence of Various Drugs for 8th-Graders, 10th-Graders, and 12th-Graders." Available at http://www.drugabuse.gov/infofacts/HSYouthtrends.html.

Cover and interior design by Annie Heckman
Edited by Mike Ingram and Jeff Link
Author photograph by Amy Werbel

Printed in the United States of America
Library of Congress Control Number: 2011924796
ISBN: 978-09840531-7-9

To United States Senator Patrick Leahy (D-Vt.)
for his years of effort in striking the proper legislative balance
between Internet safety and civil liberties.

America's future will be determined by the home and the school. The child becomes largely what it is taught, hence we must watch what we teach it, and how we live before it.

—JANE ADDAMS, America's first female winner of the Nobel Prize

What I did in my youth is hundreds of times easier today. Technology breeds crime.

—FRANK ABAGNALE, author of *Catch Me If You Can*

Contents

Part Two: The Cybertraps

Part Three: The Solutions

Part Four: Resources & Tools

Introduction

The idea for this book has been percolating in the back of my mind for some time now. As a member of the Burlington (Vt.) School Board, in the spring of 2009, I helped put together a presentation on teen "sexting"—a phenomenon born of mobile-phone technology that allows teens to send explicit images via text message. Along with a state's attorney and a detective from the Burlington Police Department, I tried to explain to board members and parents some of the thorny legal issues surrounding teen sexting cases. At the time, teens distributing nude images of themselves or their peers via text could be prosecuted under the state's child pornography laws, and a conviction could land them on the state's sex offender registry (though the Vermont legislature was working on a revision to the law that would give prosecutors some flexibility in dealing with teen sexting cases). As I researched the issue and discussed possible legislative language with my state representatives, it became clear to me that sexting was a growing problem for parents and law enforcement around the country.

That initial seed of an idea began to germinate about seven months later, during a November 2009 luncheon in New York with my friend Jeff Liebster, the managing partner of a legal recruitment firm. That day, I got a first-hand glimpse of why Jeff is so successful at what he does: with a few deft questions, he stitched together the various strands of my professional work and interests—law, technology, privacy, computer forensics, school law—and suggested (among other things) that I write a book to help parents understand the many ways their children could jeopardize their futures by doing something stupid or careless online. Did such a book exist?

After we'd paid our check and said our goodbyes, I spent some time browsing at a nearby Barnes and Noble. It was a small sample, admittedly, but I saw nothing that discussed the issue of child online safety from the perspective of children as potential perpetrators. Some additional research followed, and that winter I began to sketch an outline of the book I would write.

The Need for This Book

At first, the need for this book remained relatively abstract. I knew there wasn't anything like it on the market, and the rising tide of sexting and cyberbullying prosecutions in the news made it clear that teens were increasingly at risk of winding up in legal trouble as a result of their online activity. But it took a crazy cross-country business trip to turn an abstract need into a compelling necessity.

Thanks to my websites (www.FrederickLane.com and www.ComputerForensicsDigest.com), I get a fair number of inquiries about my computer forensics work. In early June 2010, I received an e-mail from a man in Missouri who wanted my help. He explained that his twenty-two-year-old son was being prosecuted for allegedly downloading child pornography using the peer-to-peer program LimeWire, and he was concerned that his son's attorney didn't have the expertise to adequately handle his defense.

The case had numerous practical difficulties. For starters, the son was scheduled to go to trial in two weeks, which meant there was no time to arrange for an independent forensic exam of his son's computer. There were also two camps involved in the defense: the son's attorney had actually been hired by the boy's mother and his stepfather, and they had no particular interest in supporting the father's efforts to bring in an outside expert. Notwithstanding their indifference and the limited time available, the boy's father retained me to come out to Missouri and offer whatever assistance I could to his son's attorney.

I got to Missouri about thirty-six hours before the start of trial, and spent a full day with the young man's attorney. The father's concerns were not entirely misplaced; the attorney was an older member of the bar, a solo practitioner with extensive criminal-defense experience but little or no experience dealing with computer cases (in fact, he had no computer in his office and no e-mail address). I spent the day before trial going over the available evidence with him, answering questions, and meeting with the local prosecutor. The rules that govern trials impose time limits on the discovery process and the declaration of experts, and since those deadlines had long since passed, I was limited to an advisory role.

The trial was wrenching. The young man's Internet Protocol address had been flagged by law enforcement for possible child-pornography downloads, and they'd executed a search warrant for the apartment he shared with his girlfriend. When officers analyzed the computer's hard drive, they found numerous videos and images that appeared to be child pornography. The young man was arrested and charged with possession of child pornography, a Class B felony under Missouri law that carries a maximum penalty of five years.

During trial, the prosecutor played excerpts of the videos for the jury. The content was extremely graphic and disturbing. It was difficult enough for the jury to watch, but clearly excruciating for the young man's parents, his siblings, his girlfriend, and most especially his elderly grandmother, all of whom had come to court to support him.

The defense was limited, consisting chiefly of some cross-examination of the state's forensics expert and testimony by the young man himself. He told the jury that he had no intention of downloading child pornography, and that he had no interest in such materials. He explained that he used LimeWire to search for adult pornography and that instead of downloading each search result separately, he "selected all" (Ctrl-A on a Windows computer) and then instructed LimeWire to download everything. He testified that it would often take hours for everything to download, and that he typically had no idea of what was coming into his computer. When the downloads were finished, he said, he'd go through and delete the files he didn't want to keep. When asked how it was that officers still found child pornography on his computer, he said he'd simply missed the files or hadn't gotten to them, given the large number of files he downloaded each time.

The jury was clearly skeptical; it took them less than an hour to come back with a guilty verdict. Three months later, the young man was sentenced to the maximum five years in prison.

During my brief time in Missouri, I spent a lot of time talking to the boy's father. As the father of two boys myself, neither of them much younger than his son, it was heartbreaking to see his pain, and the intense frustration that there was nothing he could do to protect his son. While conceding that his son had been careless and stupid about his online activity, he fiercely defended his innocence. And he berated himself for not knowing enough to protect his kid. "I'd never heard about this LimeWire," he said. "Why doesn't someone tell parents how easy it is for their kids to get into trouble like this?"

The Title for This Book

It was a good question, and I had a lot of time to think about it the next day as I drove from Missouri to Washington, D.C., where I was spending the summer with my partner, Amy Werbel. In the spring, Amy had received a fellowship from the Smithsonian American Art Museum to conduct archival research for her next book, *American Visual Culture During the Reign of Anthony Comstock.*

Over the years, both Amy and I have written about Comstock, a fascinating and controversial figure in American law and culture. Born in 1844 in New Canaan, Connecticut, Comstock served in the Union Army during the Civil War and, after being mustered out, moved to New York. A product of a strict Congregationalist upbringing, Comstock was horrified by the licentiousness of the big city, with its saloons, its brothels, and its active traffic in pornographic materials.

Comstock began a one-man campaign to wipe out indecent books and photographs, but the problem was vast and his efforts were stymied by a lack of resources and the indifference (if not outright opposition) of the police. He appealed to the leaders of the Young Men's Christian Association for assistance, and soon found himself an employee of the organization, charged with working to shut down pornographers by charging them with violating New York's state obscenity law. Both Comstock and the YMCA leaders soon concluded, however, that the state law was too weak to be effective.

In the winter of 1872, the YMCA sent Comstock to Washington to lobby for a new and tougher anti-obscenity law. Despite the aid of some powerful friends (including Supreme Court Justice William Strong, who actually helped Comstock draft his proposed bill), it was unclear that Congress would have time to consider the issue, since it was consumed at the time with the Crédit Mobilier of America scandal, in which numerous Congressmen had accepted bribes and shares of stock in connection with the building of the transcontinental railroad. At literally the last minute, however, Comstock's bill was passed, and on March 3, 1873, it was signed into law by President Ulysses S. Grant.

Under the new law, individuals who "knowingly deposit-ed" obscene materials in the mail could be convicted of a mis-demeanor, fined between $100 and $5,000, and could face up to ten years of hard labor.

As part of the legislative maneuvering that led to the law's passage, Comstock was appointed a special agent of the United States Postal Service and given authority to help enforce the new law, which quickly became known as the Comstock Act. A short time later, New York state passed a tougher version of its obscenity law and gave Comstock, as secretary for the New York Society for the Suppression of Vice (NYSSV), additional legal authority to seek and prosecute violators of the law.

To say that Comstock undertook his new work with en-thusiasm would be a gross understatement. He spent the next forty-two years (he died in 1915 while still a USPS special agent) vigorously pursuing all manner of potential violators: the news-boys who sold dirty pictures on the street, clerks who sold cata-logues of "life drawing illustrations" at the Art Students League, fine art purveyors like Alfred Knoedler, and even world-renown playwrights like George Bernard Shaw (who mockingly dubbed America's moralistic streak as "Comstockery").

Over the course of his career, Comstock later boasted, he arrested enough people to fill sixty-one railroad passenger cars, and was unfazed that sixteen of his prosecutorial targets committed suicide rather than confront the burly hunter of ob-scenity. He collected and burned (often before trial) hundreds of thousands of photos, books, magazines, circulars, playing cards, advertisements, newspapers, and rubber goods that he felt violated either state or federal law. He traveled endlessly around the country, advocating for state and local versions of the Comstock Act and urging the formation of local versions of the NYSSV. But by the turn of the nineteenth century, thanks in part to epithets like Shaw's and to innumerable mocking cartoons, Comstock had become more punchline than terror.

A number of factors contributed to the shift in public perception, including the desire of an increasingly globalist America
to throw off its provincial reputation and, of course, the relentless march of technology. During Comstock's lifetime, both
the camera and the movie camera made the production of "indecent" and "obscene" material so inexpensive that virtually
anyone could do it (not unlike the impact the Internet would
have almost a century later).

Though Comstock's name has largely faded from public
awareness, his legacy persists. Over the years, Congress has
adapted his obscenity law to new technologies—the telephone,
computers, and, most recently, the Internet. While some of the
law's language concerning "indecency" was eventually stripped
away by the Supreme Court as a violation of the First Amendment, its ban on obscene materials remains and is an under-
current in many of the cybertraps discussed in this book.

During his life, Comstock's main objective was to prevent children from falling victim to the immoral threats he
saw lurking in nearly every corner of the city (although he did
not hesitate to throw children in jail if he felt they were more
criminal than victim). In 1883, Comstock published a book entitled *Traps for the Young,* in which he described a variety of
threats to the innocence of youth, including "half-dime nov-
els," newspaper advertisements for alcohol and certain kinds
of "low theater," gambling, quack medicine, "free love" (mostly
prurient literature), artistic traps (nude or semi-nude classics
Comstock believed to be indecent), and liberalism. Comstock
would have a lot in common with contemporary social conservatives, and no doubt would be horrified by the Internet.

My decision to call this book *Cybertraps for the Young* is a
deliberate reference to Comstock and the book he published 128
years ago, not as an endorsement of his views but as a reflection
of the significant imprint they have left on the American legal
system, particularly as it relates to minors. In no small part be-

cause of his efforts, there are laws on the books today that are being applied to activities that could not have been anticipated five years ago, let alone fifty or a hundred years ago. There is a growing disconnect between the technical sophistication of our children and the laws that were originally intended, at least in Comstock's view, to protect them. As a result, children are finding themselves in far more painful traps than Comstock could ever have imagined.

Take, for example, the recent case of Evan Emory, a twenty-one-year-old Michigan man who has been charged with distributing child pornography and faces up to twenty years in prison. According to a recent article in the *New York Times,* Emory made a class visit to a local first-grade classroom to sing songs he'd written. He taped his performance, and after the kids left he continued taping while he sang different, sexually explicit versions of the songs, complete with hand gestures. Using a computer, he combined the footage to make it appear that he'd sung the sexually explicit songs to the kids, most of whom can be clearly identified in the video. The goal, he told the *Times* reporter, was to create a comedy sketch by pairing explicit songs with an incongruous audience. The humor came to a crashing halt when detectives seized his computer and iPhone for forensics analysis and he spent a night in jail.

Emory is nominally an adult, but what he did could have been accomplished at sixteen, fourteen, or even twelve. In fact, he apparently wrote the sexually explicit versions of the songs when he was sixteen and stashed them away until an opportunity came to use them. The technical piece of recording and combining the video images was, to put it mildly, child's play. Unlike Comstock, it's not my aim here to make moral judgments—as Emory's case makes clear, there are lots of moral gray areas—but rather to make you aware of the potential legal ramifications of your children's behavior. Some would say Emory's video was a joke that while in questionable taste did little real harm, while others would undoubtedly argue that there's nothing funny, or appropriate, about mixing sexual humor and children. But the fact remains that, whatever you think of his actions, they landed him on the wrong side of the law.

Certainly, there is a role for moral education as we prepare our children for adulthood. A product of his time and upbringing, Comstock believed that it was the role of government to provide such a moral education—to all of its citizens. He had no qualms about declaring that it was his own moral standard that should be imposed, regardless of disagreement, dissent, or diversity. The rigidity with which he stuck to that position helps explain both his initial success and his ultimate descent into ridicule.

This book is dedicated to the proposition that morality works best when it is taught from the ground up, household by household, rather than from the top down through legislative edict. It is our responsibility as parents to communicate our values to our children, including our expectations for moral behavior. That is a particularly crucial obligation in a world in which children are being given adult equipment at earlier and earlier ages.

Part One: The Technology

Chapter One
Your Child LOVES Technology

There was a time, not so long ago, when computers hid in the bowels of universities and corporations; when it took a quarter and a trip to the local arcade to play a video game; when the only way to make a phone call outside your house was in a glass booth with an outdated phone book; when a photo-processing clerk would be the first person to lay eyes on your vacation pictures.

Things have changed a bit over the last twenty-five years or so. Now, each of these activities can be accomplished with a device small enough to fit in the palm of your hand—and, increasingly, all these tasks can all be accomplished with a *single* device.

Our children are growing up in a world that is awash in remarkable digital technology. Kids seem astonishingly well suited to this world, as they effortlessly navigate complicated menus, rapidly master new programs, and nimbly text, tilt, and click away. It's no wonder that post-Web/Millennial children are often referred to as "digital natives," implying (correctly) that the rest of us are just strangers in an increasingly strange land.

Statistics offer some insight into just how much kids like technology. According to a recent study by NPD Group, a market researcher of technology trends, 82 percent of kids between the ages of two and seventeen—55.7 million U.S.

children—describe themselves as "video gamers." That includes a surprising number of kids who may not even be able to read the survey questions. All told, 9.7 million children between the ages of two and five reportedly play video games.

The adoption of other technology by kids is just as enthusiastic—and just as startling. Nearly four years ago, the Kaiser Family Foundation found that roughly a quarter of all children between the ages of four and six were using personal computers at least fifty minutes a day. Amazon.com offers more than twenty-five digital cameras designed specifically for children aged two to four, and the average age for a child's first mobile phone is now under ten.

On their own, these cool technologies pose few legal risks to children (although they may have other effects, such as increasing distractibility and decreasing exercise—topics for another book). The real problems arise with the three C's of technology: communication, capability, and convergence. More and more often, we're handing our children remarkably powerful devices long before they have the wisdom or maturity to understand the consequences of misusing them.

Children are using computers and mobile phones to bully and harass each other. They're using the cameras on their phones to take nude photos of themselves and others, and to send those photos to dozens or hundreds of other children. They're using a variety of electronic devices to cheat in school, steal intellectual property, and commit a wide variety of crimes, from identity theft to hacking.

In short, it has never been more important for parents to take the time to understand how every device works, to think through the consequences of giving these devices to their children— regardless of their age—and to talk with their kids about how to use their electronic gadgets responsibly.

Mobile Phones and Smartphones

In terms of capability and communications, no category of consumer device has changed as dramatically as the mobile phone. First introduced in the United States in 1983, early handheld phones were widely referred to as "bricks," and did nothing more than allow people to free themselves from fixed phone locations (though that seemed amazing enough at the time). Around the turn of the twenty-first century, the development of faster cellular networks (first 2G, then 3G) allowed for the transmission and downloading of multimedia content. Ringtones were the first media content to be widely distributed, followed quickly by games, photos, and eventually even streaming video.

The idea of the mobile phone as a connectivity tool began to creep into mainstream consciousness in 2003, when Motorola Inc. released its hugely popular Razr phone in the U.S. The sleek, slim, clamshell phone was equipped with a low-resolution camera, a 2.2-inch LCD screen, and various communication options, including text messaging and a simple Web browser that could be used to send e-mail. In many ways, it was the first device that was as much a portable tool for surfing the Web as it was a phone (in fact, during the height of the phone's popularity, the Web browser Opera released a Razr-specific version of its software, which offered more features than the phone's own browser).

In the seven years since the Razr was released, the percentage of kids using mobile phones has risen from twenty to approximately ninety-five. Part of that growth was the result of a conscious effort on the part of mobile-phone companies, beginning in 2004, to target advertisements at the teen market, a move that was obviously highly successful. Another factor was Motorola's success in making the mobile phone a fashion item—the Razr not only looked cool, but came in a variety of attractive colors. But it was SMS messaging, or "texting," which did the most to'bring teens and mobile phones together. More recent billing data is still being analyzed, but between the first quarter of 2006 and the second quarter of 2008, the number of text messages sent in the U.S. skyrocketed from 65 million to 357 million. It's not surprising, then, that in October 2010, 43 percent of teen mobile-phone users reported that their primary reason for having a phone was to send texts to friends. The SMS feature was the most frequently cited benefit of phone ownership, with "safety" and "keeping in touch with friends" a distant second and third, respectively.

Games and Web surfing didn't even make the list of top reasons for owning a mobile phone, but it's likely that will soon change. Apple Inc. ushered in the real era of handheld computing and surfing with its release of the iPhone on January 9, 2007. What makes the iPhone so remarkable—and so potentially troublesome for parents—is its seductive combination of well-designed and powerful hardware, flexible software, and wireless connectivity. It has been a tremendous hit.

In early 2008, Apple announced the creation of the iTunes Store, which allowed programmers to sell their own applications ("apps") for use on the iPhone. There are now hundreds of thousands of apps for sale, and iPhone users have logged well over a billion downloads in the three years since. Although Apple has aggressively policed the iTunes Store to prevent the sale of obscene, indecent, and even politically provocative apps, there are still hundreds, even thousands, of iPhone (or Android or Blackberry) applications that can land kids in trouble. For instance, every major social network site—Facebook, MySpace, Twitter, etc.—has its own app for posting photos or comments, and dozens of third-party apps offer additional tools for interacting with those sites. There are also hundreds of apps—Pixelpipe, Instagram, and so on—that are specifically designed to make it easier to take and upload photos to social networking sites.

The enormous popularity of the iPhone has fueled a mobile computing arms race. Hardly a month goes by without a new phone or mobile operating system hitting the streets. A year after the iPhone debuted, for instance, a consortium of companies (including Google) announced the release of Android, an open-source mobile operating system. Other manufacturers, including Research in Motion Ltd. (RIM), Nokia Corporation, and Microsoft Corporation, have announced the release of updated versions of their mobile operating systems, along with their own platform-specific app stores. Not to be outdone, Apple has released an upgraded version of its iPhone on a yearly basis; the most recent model, the iPhone 4, allows users to conduct face-to-face video conversations using "Face-Time" (a feature which the adult entertainment industry is already exploiting for pornographic purposes).

Right now, only 23 percent of U.S. teens have a smartphone such as an iPhone or Android, but that percentage will no doubt climb quickly.

Videogame Consoles

It would be difficult to imagine someone less likely to be a video game fan than my grandfather, a serious and well-respected real estate lawyer who spent more than fifty years at the forefront of his legal specialty. I remember debating him when I was in law school about the merits of computer-aided legal research, a development he disdained. But even Grum could feel the lure of electronic gadgetry: in 1975, he was one of the early purchasers of Pong, Atari Inc.'s groundbreaking video paddle game. I was ecstatic: I'd never seen anything like it, and happily spent hours playing the game on my uncle's ancient, black-and-white TV.

Regardless of how long I played Pong, my parents had little reason to worry about what I was doing—the Pong console played Pong and only Pong. But video game consoles have gone through a few changes since then. Today, the capabilities of those consoles rival or exceed the capabilities of most desktop and laptop computers (in fact, law enforcement agencies use the latest game consoles to help crack passwords). Recent models have incorporated Blu-Ray disc players, motion-sensing and wireless controllers, and high-end video processors for increasingly lifelike and immersive game-playing experiences (which helps explain why 10 percent of teens log more than twenty hours a week playing video games).

In part because of those new features, the devices are incredibly popular with children (and, no doubt, with their parents as well). In early 2010, the Pew Internet and American Life Project reported that 81 percent of children between the ages of twelve and seventeen own at least one gaming console, while 51 percent own a portable gaming device—a figure that's almost certainly risen over the past year, thanks to the popularity of game-capable devices like the iPhone and other late-generation smartphones. The combined sale of video game software and hardware generates over $20 billion per year in the United States alone.

The most significant development, however, is the growing connectivity of these gaming devices. All three major consoles—Sony Corporation's PlayStation 3, Microsoft's Xbox 360, and the Nintendo Co., Ltd.'s Wii—are designed to connect directly to the Internet. Gaming companies added connectivity to video consoles to encourage gamers to buy new games online and to compete with opponents playing the same games around the world. Most contemporary games also have built-in messaging capability, so gamers can communicate with each other in real time.

Keep in mind that the same connectivity, combined with the rapidly expanding features of gaming console software, allows gamers to do everything online that they have traditionally done using a computer—send e-mail, use Internet chat and instant-messaging services, surf the Web, and send and view photographs. The potential dark side of gaming consoles was unveiled last year, when a Kentucky man was arrested for using the communication capabilities of his Sony Playstation 3 to strike up a relationship with an eleven-year-old girl and solicit nude photos from her online.

Not long ago, a portable gaming device would have been a safer alternative—the only thing handheld consoles were designed to do was play games. Like so many other types of technology, however, recent models of handheld consoles— the Nintendo DS, the Sony PlayStation Portable (PSP), and the Apple iPhone and iPad—are designed to connect to the Internet for game downloads, interactive play, and messaging. In addition, both the Nintendo DS and iPhone are equipped with built-in cameras.

Digital Cameras

Many industries have legitimate reasons to bemoan the rise of digital technology, and photo processors are near the front of that line. Not so long ago, it seemed as if every shopping mall was equipped with a drive-up Fotomat kiosk offering one-day film processing. But Fotomat didn't survive the rise of one-hour developers in malls and box stores; in turn, those one-hour shops have largely been replaced by drug-store kiosks— or even inexpensive home printers—that can produce prints directly from CDs, USB sticks, or memory chips.

The culprit for all this retail change is the digital camera, a device that first became commercially available only about twenty years ago. As prices have steadily fallen into the low double digits, digital cameras have become ubiquitous—106 million were shipped in 2009 alone, and that figure was actually down about 12 percent from the 119 million units sold the previous year (a drop attributed in large part to the growing popularity of mobile phones equipped with cameras). The attractions of digital cameras are obvious: the ability to see photos as soon as they're taken, limited costs for film development, and software that makes it easy to save images to a computer, digitally manipulate them, and instantly distribute them via e-mail, a blog, a website, or a social network.

As the Amazon.com catalog illustrates, digital cameras are often the first electronic devices to be put into small hands. There are a number of benefits to doing so: photography can be a terrifically rewarding hobby, and, by eliminating the costs of film and film processing, digital cameras have helped thousands of children discover a new artistic medium, a new way of looking at the world around them. It's the very simplicity and ease of digital cameras, however, that have turned them into increasingly insidious cybertraps. To paraphrase the old saying: just because a photograph *can* be taken doesn't mean it *should* be taken.

As kids get older and start exploring their sexuality, they all too often employ digital cameras or mobile-phone cameras as part of that exploration. Few fully understand, however, that taking nude or semi-nude photographs or video of someone under the age of eighteen (even a self-portrait) can have serious, even life-altering, consequences. As I'll discuss in more detail in chapter eleven, children who have circulated these types of photographs have been subjected to cyberharassment and cyberbullying laws, have been suspended and/or expelled from school, have been prosecuted under state and federal law for producing and distributing child pornography, and, in a few extreme cases, have committed suicide.

To make matters worse, digital cameras are steadily shrinking in size, making it dangerously easy to take photos or video of someone without their knowledge or consent. In fact, there are entire websites devoted to illicit candids conspicuously labeled as "upskirt" or "downblouse."

Even if the subject of a photo is aware he or she is being photographed, the speed and simplicity of digital technology reduces inhibitions and makes it less likely that either the photographer or the subject will consider issues like privacy or propriety. And the younger the kids on either side of the camera, the less likely it is they'll think about the long-term consequences of what they're doing. After all, there's no longer any risk for your child that he or she will feel embarrassed when you pick up that packet of racy photos at the developer; there's the illusion that what's taken with the camera will stay in the camera.

But that sense of privacy truly is an illusion. The whole point of a digital camera, after all, is that the images can easily be transferred to a computer, thus freeing them for distribution to the rest of the world. More recent digital cameras skip the transfer process altogether—they're equipped with wireless transmitters that allow photographs to be sent directly to e-mail or social networking sites. In fact, there's even a memory card, the Eye-Fi, that will enable older digital cameras to transmit photos wirelessly. Not to mention that any smartphone equipped with a camera (and most are) is specifically designed to transmit photos in a variety of ways—as e-mail attachments, as MMS messages, or as direct uploads to any of hundreds of websites and social networking services.

Desktops, Laptops, and Tablets

A decade into the twenty-first century, it's impossible to overestimate just how important computer skills will be to our children and grandchildren throughout their lifetimes. Consider the fact that people born when the first personal computer was released in 1978 are already in their early 30s, many with children of their own. Computers are an integral part of our education, our work, and our leisure, so much so that the research firm NPD Group estimates that 97 percent of American households own a computer—a level of ownership that rivals televisions and stereos.

The importance of personal computers was underscored for me when, while writing this book, I spent a day in the reading room of the Widener Library at Harvard University. The long wooden tables were filled with undergraduates, graduate students, and researchers, and nearly all of them had laptop computers open on the table in front of them. Assuming an average value of $1,000 per computer (the numerous Macs in the room bumped up the average a bit), those library tables were covered with at least $100,000 in portable computing equipment. As a society, we've invested a staggering amount of money in hardware and software, and it would be unrealistic to think that kids can succeed in school and in the workplace without at least some familiarity with and exposure to this technology.

For parents, however, there's a tension between the obvious benefits of computer ownership—homework, research, entertainment, socialization—and the distractions and dangers that come with it. Thanks to the development of the World Wide Web in the mid-1990s, every computer is now a portal to an infinite array of information, entertainment, and enticements, not all of which are salutary.

As we'll see throughout this book, many kids need nothing more than a keyboard and an Internet connection to get themselves into a world of electronic trouble. Computer hacking, for instance, has a long, albeit disreputable history, and enterprising teens have often been at the forefront of the mischief. But even less computer-savvy teens now have at their fingertips all the tools they need to violate a host of laws. With any type of Internet-connected device—laptop, mobile phone, gaming console—children can harass or bully other children, libel their teachers, commit fraud on eBay, violate U.S. copyright laws, or commit a felony by downloading or distributing obscene materials. Given the prevalence of computer web cams—either clip-ons for desktops or pre-installed in laptops—kids have the ability to broadcast themselves and others from the privacy of their bedrooms to a global audience, and that behavior has the potential to violate any number of federal and state laws.

We don't normally think of desktop or laptop computers as phones, but thanks to web cams and online conferencing software like Skype, they certainly can be used that way. Actually, the line between computers and handheld devices is blurring in both directions. Kids can use computers like phones (albeit bulky ones) through the use of software like Skype and Google Talk, while mobile smartphones allow them to perform a number of computing functions—e-mail, Web browsing, photo sharing—on the go.

In fact, it is only a matter of time before such handheld devices match traditional computers in both power and convenience. A vision of that future could be seen in April 2010, when Apple announced the release of its tablet computer, the iPad. There have been numerous previous attempts at creating and marketing a tablet computer, but with its inimitable attention to detail, design aesthetic, and sheer coolness, Apple succeeded where so many others have failed. During the 2010 holiday season, online retailers and tech journalists reported that the iPad was squarely at the top of the teen wish list, despite its impressive price tag.

Right now, the iPad is an imperfect bridge between laptops and the iPhone. Out of the box, it lacks a physical keyboard, has no camera, and has relatively limited onboard storage. Both Apple and various third-party vendors, however, sell keyboards that dock onto the iPad; future versions of the iPad are likely to include both a rear- and front-facing camera (for "Face Time" chat, among other uses), along with more memory and faster processors.

The iPad's enormous success—more than 25 million sold in fiscal year 2010, with another 32 million projected for FY 2011—has attracted numerous competitors, including the Samsung Galaxy Tab, the HP Slate 500, the Dell Streak, and the Motorola Xoom. How those tablets will fare in the marketplace remains to be seen, particularly with Apple's iPad 2 waiting in the wings, but one thing is clear: computing's own Olympic motto—smaller, faster, cheaper—will continue to drive developments in the tech world.

With relatively few exceptions, these new electronic devices and the ones that inevitably follow will offer more options, more capabilities, and more ways for kids to get into trouble. And kids will continue to love them. This book aims to level the playing field a bit, to give parents and teachers a solid introduction to the ways in which kids can accidentally or intentionally misuse electronic gadgets. I'll begin with a general overview of the communication revolution, describe the potentially life-altering cybertraps that can snare kids, and then offer some suggestions on how to protect your child.

Chapter Two
A Parent's Guide to the Communication Revolution

For many parents, the idea of supervising their children's online activity can seem overwhelming. Computer and mobile technology often is bewildering in its complexity, and the phenomenal pace of change makes it all the more daunting to stay on top of the latest innovations. Add in the stress of multiple jobs, multiple kids, multiple pets, not to mention all the other distractions of modern life, and it's understandable why teens and preteens are frequently left (literally) to their own devices.

There's no question that keeping up with all the gadgets, software, and websites designed specifically to fascinate children can turn into a full-time job. Fortunately, it's not necessary to be an IT specialist to protect your child online, and this book isn't intended to turn you into one. What this book *is* designed to do is help you identify the basic information you'll need in order to guide your children's behavior and minimize the risks—including serious legal risks—they can incur in their online lives.

Regardless of your personal level of tech savvy, there are several simple, straightforward questions you should always ask before buying your child the latest electronic gadget:

1. What types of information can the device or software collect or distribute?
2. Can it be used to communicate with others, and if so, how?

3. How much data does it store, and where?
4. Can your child change the device's capabilities without your knowledge?
5. Can you monitor your child's use of the device, and if so, how?

As a parent, the emphasis should be less on how your child is interacting with a specific program or device, and more on how your child is interacting with other people, whether online or off. Are they respecting personal property and the rights of others? Are they acting as a bully? Are they violating someone else's privacy, or their own?

Kids are remarkably effective at making their parents feel as if they're out of touch, particularly when it comes to technology. More often than not, it's a semi-conscious strategy designed to keep parents off balance and at arm's length. But what's often overlooked by parents—and consistently by children—is that with age comes experience. The tools of human interaction may change rapidly, but the social and legal consequences change far more slowly, if at all. Parents are uniquely positioned, regardless of how technologically adept their children may be, to help guide children through those sticky thickets.

Equipment Capabilities

Unfortunately, your life experience alone is not enough to guide your children through the risks of technology; it's also necessary to have at least a basic grasp of the technology itself. The single most important question parents need to ask about any electronic device is, what can it do? More specifically: what types of information can the device create, collect, and distribute? The answer to this fundamental question will help determine what types of cybertraps your child might face.

A great place to start would be to ask the person selling you the device about its capabilities. You also could consult with a friend or family member who's already using one. A third option is to talk with the child to whom you're giving the device. Kids are often remarkably tech-savvy, and asking your child how they plan to use a new electronic gadget could be a great bonding opportunity, as well as a chance to start a conversation about appropriate behavior. Keep in mind, of course, that children may not *want* you to know everything they can—or will—do with a particular device; it's important to conduct some independent research as well.

The other thing to keep in mind is that, increasingly, the answer to the question, "What can this device do?" will be some variation of "everything." As we discussed in the first chapter, there is a growing trend toward convergence—the tendency of all devices to have similar data-capture and communication capabilities, regardless of their primary function or manufacturer. As a result, it's probably safe to assume until proven otherwise that any recent-model electronic device is capable of capturing and distributing just about any information your child desires.

For instance, most recent laptop computers come equipped with a webcam that allows users to take photographs and video, which can then be uploaded to a social networking site or sent via e-mail to an endless number of people. The same can be done with any recent-model smartphone, or several of the new tablet devices that run Apple's iOS or Google's Android operating systems. Gaming consoles, too, can be used to access the Internet, exchange instant messages, chat, send e-mails, or upload and download files. A simple digital camera may seem like a safe bet, until you notice that many recent models are equipped with wireless Internet capability and built-in software for uploading photos directly to social networking sites like Facebook.

Communication

The next critical question to ask about a device is, What can your child do with the information (text, photographs, video) that he or she creates or captures with it? For kids, it's all about communication—how effectively (and coolly) can a particular device be used to interact with other kids?

1. Texting (SMS and MMS)

For much of the computer era, e-mail has been the killer communication app—so much so, in fact, that there's a faint but growing chalk outline around the U.S. Postal Service. Kids, however, find e-mail as hopelessly archaic and slow as their parents find handwritten letters. And, to kids, actual phone conversations aren't much better. For this generation, the communication tool of choice is texting.

Texting currently comes in one of two flavors: short messaging service (SMS), which is limited to actual text messages of up to 160 characters in length, and multimedia messaging service (MMS), which allows mobile-phone users to send messages containing both text and media content (photos or videos). Most wireless services also allow users to send and receive text messages through their e-mail accounts. As you look at different phone models, you'll soon see that although each uses slightly different software for sending text messages, the basic concept is the same: pick a recipient from a contact list or enter a mobile-phone number, type a message, and press "send." Teens can grow remarkably adept at texting; it's not uncommon, for instance, for kids to text in the dark or without even looking at the keyboard.

For anyone over the age of twenty, it can be difficult to fully grasp just how enthusiastically the current Millenial generation has embraced the practice of texting. A recent Nielsen study reported that the average teenage boy sent 2,400 text messages every single month. The average girl out-texts the average boy nearly 2 to 1, sending just over 4,000 texts per month—or roughly one text every ten minutes. In fact, texting has become such an integral part of a child's communication repertoire that child-development specialists have begun to study its importance in preteen and teen social development.

It's a trend that's likely to continue, as social networking sites like Facebook begin to add text-like capabilities to their services. In a recent interview, Facebook's director of engineering, Andrew Bosworth, told the *New York Times* that "[t]he future of messaging is more real-time, more conversational and more casual." To embrace that future, Facebook has already changed its messaging service to eliminate the subject line, remove the "cc/bcc" options, and turn the "enter" key into a "send" button.

As we'll see in chapter 6, the ease, speed, and lack of face-to-face contact inherent in texting has contributed to some of the behavioral problems that can turn into cybertraps—most notably harassment, cyberbullying, and sexting (the practice of sending text messages with nude or semi-nude photos). Part of the trouble is that texting—like so much other online behavior—can feel private, or even anonymous. But that privacy, of course, is a deception. One of the hidden truths of the electronic age is that a large amount of activity can be tracked across the Internet, and even more electronic activity can be retrieved and reconstructed from the devices used to generate it.

If your child has the ability to send SMS and MMS messages by phone (and the odds are overwhelming that he does), then it's already past time to have a conversation about the moral and ethical boundaries of electronic communication. What's important is not the specific software your child is using to send messages, but the content of the messages themselves—what your child is saying, and/or showing, to others.

2. Instant messaging
Another popular form of communication among kids is instant messaging. It's similar to texting, but without the 160-character limit. As its name implies, instant messaging allows indi-

viduals to send messages back and forth in real time. It's analogous to face-to-face conversation, except it takes place entirely online.

Instant messaging first gained popularity with AOL Inc.'s AOL Instant Messenger (AIM) service, and quickly spread to other services like MySpace, Gmail, and Facebook. Typically, these systems are designed to allow individuals to communicate one to one, but a number of services are expanding to allow group messaging (essentially an individualized form of online chat).

Even if a service doesn't allow group messaging, kids can create an ad hoc version of group chat by opening multiple messaging windows with various friends. I've seen the twelve-year-old in my house simultaneously instant-messaging with four, five, six, or even more classmates through his Gmail account.

As with texting, there's nothing inherently wrong with instant messaging. As we'll see in chapter six, potential problems arise not from the fact that kids are instant messaging with each other, but from what they're saying *in* those messages.

3. Online chat

Online chat is the group version of instant messaging. It's by far the most notorious of the various ways children communicate online, since it plays such a frequent role in child solicitation and molestation cases. To give you a flavor of what the online chat conversations can be like, here's a brief sample taken word-for-word from one teen-oriented chat room:

<NEWPORT13> ahh any girls wanna a chat16f...Im a stud lookin for cute girls to chat with
<flyleaf2010> is cute
<flyleaf2010> kid
<troy788> TROY SAYS, BEING GAY IS A DISEASE
<bhsathlete> delia
<LoveAlwaysDee> IM GETTING SICK OF DUDES MESSAGING ME . WHAT PART OF IM GAY DONT YOU GET ?!!!

<xxdashaxx> hey 16 male australia here:-)
<Relentless-> :D
<tasteslikelemons> delia?
<bhsathlete> hey
<flyleaf2010> wha
<flyleaf2010> hi
<flyleaf2010> person?
<onedreamoneplace> hey 17 year old male from so cali(sandiego actually) im sweet, caring, honest, and not a perv or freak. im reall chill, and outgoing, and funny (so ive been told) lookin to get to know locals and new friends and meet genuine people so hit me up:)
<bannedWTF> 16 F LOOKING FOR GIRLS WHO R USED BY PARENT(S) P2P ME
<rynstorey18> Male gay 17 hit me up :)
<tasteslikelemons> hai
<ayyitssmee> BEING TROY IS A DISEASE !
<betterandbetter> I don't show that much emotion at any time,just some things fuck me in the head.
<flyleaf2010> hey
<magickc1997> any cute guys wanna chat?*any cute guys wanna chat?
<troy788> BEING TROY IS A BLESSING
<silence_413> any guys from socali wanna chat? 17/m/cali
<tasteslikelemons> does the old username MercuryFreak ring a bell?
<jonas_usa> 16 Male BRA Bored and Single.!! But Just Looking for new Friendships** =D xoxo ADD+
<Starfish_Loves_You> YOUR ANUS IS A DISEASE STFU
<ayyitssmee> IN WHAT WORLD HUN?
<Starfish_Loves_You> >_>
<ikay_kay> 14 f bi p2p me if you wanna chat =]
<gayboy41> hey ppl
<flyleaf2010> yes
<betterandbetter> Yes it does Lemon
<lezlisa56> ne bi's or lesbians wanna chat p2p me now
<natedawg1117> guy looking for girl or guy message me
<Ciesliko3> dasha p2p me
<OmgitsDee> being troy is a disease hey Troy when did u get the Straight disease?
<mevio5> 14/f/conneticut baby! bi..GET AT MEEE! (:
<Christina11811> female bi 13 pm me ;)
<tasteslikelemons> this mahh new account
<flyleaf2010> forgot your name
<MrsPrincess> 17 female texas bi lookin for conversation, friends, n txtin buddies hit me up clean chat only NO ASL NO CHAT
<Kdavis93> BEING GAY IS A WAY OF LIFE IF U DONT LIKE IT SO BE IT U

DONT HAVE TO COME ON HERE AND MAKE FUN OF EVERY ONE HERE
<LostRomantic17> 17/f/indiana, any cute girls wanna chat? pm me
<gayboy41> hey everyone
<gayteenct> gay boy ct area under 18
<ieshia23> i need a gf
<betterandbetter> Hi Gay
<tasteslikelemons> isnt that cute. my feelings are hurt
<flyleaf2010> merci???
<troy788> TROY IS CLEAN DEE, NO DISEASE HERE
<akl93> 17m, anyone 17 or older hmu :)

Keep in mind that the chat room in question had from 150 to 160 people in it during this conversation, and the lines above appeared on the screen in little over a minute's time. Whatever kids are doing online with their electronic devices, they're doing it at the speed of light. And clearly they're sharing information, using language that would make a sailor blush, and all too interested in taking the conversation (and the proposed activities) offline. An unknown percentage of the anonymous chatters are undoubtedly men of a certain age, and at least as many are law enforcement officers, but this is likely not your parents' playground (or yours, for that matter).

The granddaddy of online chat is a service known as Internet Relay Chat, or IRC. In order to access IRC groups, a user has to first install chat software, the most common examples of which are mIRC, Xchat, and Chatzilla (for a more extensive list of IRC clients and program names to look for in your own computer, visit the book's website at www.cybertrapsfortheyoung.com). Chat software makes it possible to log into any one of thousands of so-called chat channels, each organized around a particular theme or topic. The Web site SearchIRC.com lists hundreds of chat channels organized by topic, including Age Groups > Teen (#TeenLounge, #teen, #antisocial, #teens), Sexuality > Chat (#sweettalk, #anything_goes, #social-kink), and plenty more, on just about any topic imaginable.

Chat rooms were one of the earliest and most popular features offered by the groundbreaking online services like CompuServe, Prodigy, and The Source, all of which helped introduce people to the idea of online interaction. Due to concerns over safety and the growth of social networking sites like MySpace and Facebook, online service providers have been steadily shutting down their chat rooms. Currently, Yahoo! Inc. is the only major online provider to still offer the feature. However, there are a variety of websites that operate specialized chat rooms, including many aimed specifically at teens and preteens: TeenChat.com, TeenFlirt.com, and TeenSpot.com, for instance.

In and of themselves, chat rooms are not intrinsically harmful. They can offer entertaining interaction with interesting people and potentially useful information for teens who might find it difficult to obtain it in other ways. For instance, TeenSpot.com offers chat rooms like "Alternative," (a chat room for gay, lesbian, and bisexual teens), "Christian" (discussion for teenagers of the Christian faith), "Advice" (for giving or receiving advice from peers), and "Dorm" (a chat room for older teenagers who prefer mature conversations).

Where trouble can quickly arise, however, is when teens use chat to harass or bully other children, or to share photos or videos of themselves or others. It is also not uncommon for people to use chat rooms as a meeting place to share copyrighted content, hacking information, or contraband material like child pornography.

As I said earlier, law enforcement pays close attention to chat rooms specifically targeted at teens, or with names suggestive of teen sexuality. Many prosecutions have arisen out of chat room conversations in which a potential predator thinks he's talking to a fourteen-year-old girl but is, in fact, chatting with a forty-two-year-old male police officer with good online acting skills.

4. Social Networking Sites, Blog Posts, and Message Groups
The final major category of online communication for kids is made up of "status updates" and "wall posts," the brief comments they post on social networking sites to let their friends know what they're thinking or doing. These posts can be stand-alone items, or they can spark long-running conversations filled with juvenile jokes, insults, bad language, non sequitors, and other drivel from teenagers who do not self-censor. Kids can attach links to these updates—to websites, photos, or video clips—and can even let their friends know if they "like" a particular status update or comment (at least on Facebook). There's also the endless potential for embarrassment through the posting of embarrassing photos of one-self or others—scantily dressed, engaged in public displays of affection with the wrong person, doing something illegal or prohibited at a party, or perhaps just being somewhere they're not supposed to be.

By an overwhelming margin, Facebook is the most popular site for these types of "microblog" posts. So far, Facebook's phenomenal popularity (more than 500 million users and counting) and its frequent feature updates have kept kids interested. But it's important to realize there are dozens, if not hundreds, of other socially oriented websites and mobile services designed to attract kids and encourage them to post content. Tracking the social networking sites children visit or the mobile services they use is probably the most difficult type of communication to monitor, at least without software assistance. With the exception of Facebook, kids are notoriously fickle about the sites and services they use, and they tend to flit quickly from one to another.

Depending on your child's level of sophistication, he or she may run a stand-alone blog on a topic of interest. It's remarkably easy to start one and then post regular updates. It only takes a few minutes, for instance, to set up a hosted blog on a site like Blogger or Wordpress and launch a blog on any subject in the world. Some of the Internet's greatest features—its openness, its low cost of access, and its reach—are precisely the ones that can lure your child into a variety of communication-based cybertraps, including cyberbullying of other children, harassment of kids or teachers, and even defamation.

Similarly, your child might be posting messages on various topics to Yahoo! Groups or Google Groups. These are somewhat less well-known Internet resources, and generally less popular with teens, since they lack the immediacy, speed, and multimedia capability of other types of online communication. Still, it's good to know what's available. Groups are essentially electronic bulletin boards where Internet users can freely and easily post messages about almost any topic under the sun. All of the communication cybertraps discussed in this book can be committed through group postings, but the odds are relatively low.

Talking to your children about what they're doing online is still the most important preventative measure you can take. That conversation should include some frank talk about the consequences their online actions could have. And if you have active concerns about your child's online activity, then installing monitoring software may be a necessary step. The possible need for surveillance software is discussed throughout the book, but for specific ideas and suggestions, see appendix A.

Data Storage

Thanks to the combination of computers and digital cameras, we're in a period of unprecedented information creation. In 2010 alone, scientists estimated that humans created 1.2 zettabytes of information. A zettabyte is a *very* large number, equal to one sextillion bytes of data. When you start talking about that amount of data, efforts to create illustrative examples get a bit ludicrous. Someone calculated, for instance, that it would take seventy-five billion iPads to store that much data (a number which might give Apple CEO Steve Jobs cause to smile).

Much of the information kids create is ephemeral—the vast majority of texts, for instance, fly from one device to another, get deleted, and are then gone for good. But lots of data and bits of information (photos, videos, status updates, blog posts) get stored in various locations. In fact, given the nature of electronic data, a single piece of information is frequently stored on multiple devices at once—which is one of the reasons it can be so difficult to get rid of electronic data once it's created.

If you're concerned about the information your child may be creating, there are generally three places where it will be stored: local storage, a mobile device, or "in the cloud." How easy it is to locate that information or data depends on a number of factors, including your familiarity with computers and software, the particular sites or services your child is using, and how much effort he or she has put into hiding it. It's important to be aware of what storage options your child has, in part to understand what he or she might be doing, and in part to understand what might be seized by law enforcement during a criminal investigation.

1. Local Storage

Local storage refers to the hard drives contained on your child's laptop or desktop computer, or to the myriad types of removable media available for home use: external hard drives, USB or thumb drives, CDs, DVDs, and so on. By and large, these devices are used for storing actual files, like photos, videos, word-processing documents, and so on. If you're concerned about photos your child has taken or material he or she has downloaded from the Internet, those are the places to begin looking.

2. Mobile Devices

Every mobile device marketed today—cameras, music players, smartphones—has some amount of on-board storage for data and information. Smartphones and music players generally have the most built-in storage capacity, typically between eight gigabytes and sixteen gigabytes, though some devices have much more. For instance, Apple offers a 160-gigabyte version of its classic iPod Touch, which the company says is enough room to store twenty-five thousand photos or two hundred hours of video.

Cameras and video cameras tend to have less on-board storage, but they're frequently equipped with slots designed for Secure Digital (SD) memory cards, which can store anywhere from 1 gigabyte to 64 gigabytes of data. Many laptop computers are designed with an SD card reader, which allows photographers to take the card from the camera, slip it into a computer, then quickly transfer the photo files. Once on the computer, the photos can easily be manipulated, uploaded to social networking sites, or e-mailed to friends and foes alike.

Small memory storage devices like USB drives and SD cards tend to multiply, so there may be a number of them lurking in cluttered corners of your child's room (or elsewhere in the house, for that matter). USB drives also show up in a variety of shapes—key chains, pens, action figures, NBA team logos, fake sushi, you name it. Frankly, the only limit to data storage possibilities are the imaginations of marketers and gag gift designers.

3. Cloud Storage

As amusing as a Lebron James-shaped USB drive may be, the real future of data storage is "the cloud," a nebulous description for services that operate exclusively on the Internet. In the last couple years, a huge number of sites have been launched that offer remote storage of increasingly large amounts of data. The advantages are numerous: essentially unlimited space, much more effective backup, access to content from any Internet-connected device, easy distribution options, and no software upgrades. In fact, Google Inc. is in the process of developing a netbook computer that will have no local storage at all; everything will be done online and stored "in the cloud."

Until such "cloudbooks" catch on, there will be a transition period in which data is generally stored on local devices, but backed up or shared in the cloud. The rapid rise of cloud storage services like DropBox, Box.net, Google Docs, Picasa, and Snapfish is complicating investigations by law enforcement and unquestionably making it more difficult for parents to monitor what their children are doing with their electronic devices.

If you have concerns about what information your child is storing online, the first challenge is figuring out what sites or services are being used, then familiarizing yourself with how they work. Keep in mind that in most cases, even if you identify an online storage site your child is using, you probably won't be able to see what is stored there without your child's username and password.

Software and Hardware Upgrades

There's something very comforting about giving a child a set of blocks as a gift. It's a tactile present, one that stimulates the imagination and feeds the inner architect. Even better: once a block, always a block (unless it becomes a missile, which is a different sort of discipline issue). You don't need to upgrade a wooden block, or add new features to it; it is what it is.

The electronic devices now popular with kids could not be more different. Computers, smart phones, tablets—all these devices are designed specifically to run new programs, accept various hardware upgrades, and power a seemingly endless number of peripheral devices. It is reasonably safe to say that few, if any, electronic devices remain in the same state they were in when first taken out of the box.

Your child will likely spend a lot of time tweaking and upgrading their computers and gadgets. You may have purchased a device with the idea that it would help your daughter do X, and now may be startled to learn that within minutes, she's figured out how to make it do Y, Z, and occasionally, even α and ω.

Effectively protecting your child requires spending enough time with a device to understand what new capabilities may have been added since its original purchase. For instance, spend some time playing with your child's smartphone or tablet, and ask your child what the various applications and programs can do. Pay attention to what has changed, particularly with respect to programs specifically designed for communication or social networking. Familiarize yourself with the programs that have been installed on your child's computer, and track the changes that have occurred. In particular, be thoughtful about hardware purchases: if your child wants to add a web cam to a laptop, or is insistent he needs a smartphone with a camera, take the time to talk about why he wants these technologies, how they will be used, and what some of the risks might be.

Monitoring Possibilities

So we come back to the practical question: given all the devices, software, mobile apps, and websites available to children these days, how can parents possibly keep track of it all? Short of surgically attaching yourself to your child (not generally a good idea), it's impossible to know *everything* they're doing online. You may be tempted to raise your children in an electronics-free zone, but that's not overly practical either, and in the long run you'd be doing a huge disservice to children who will have to function in a digital universe for the rest of their lives.

How much you really need to know about your child's daily online activity is a calculation based on your child's maturity, the maturity of his or her friends, the amount of free time he or she has (idle hands are the devil's plaything), whether you've witnessed any disturbing changes in his or her behavior, and your personal familiarity with technology. Fortunately, despite the myriad electronic options for your child, the forms of communication most likely to cause problems can be monitored.

For instance, if you're worried about what your child could be saying or showing in text messages, you can ask your mobile-phone company to provide you with copies of all the texts your child has sent or received in the last month. Keep in mind, however, that this could involve an awful lot of reading.

Unlike text messages, there's no intrinsic means for parents to monitor instant messaging conversations as they're taking place. However, some services that offer instant messaging also provide the capability to record instant messaging conversations. Gmail, for instance, has built-in instant messaging that can store conversations in the same way e-mail conversations are stored. On this book's website, there is a page that lists the most popular instant messaging tools and whether the conversations can be electronically recorded.

Like instant messaging, IRC clients generally do not record conversations that take place in a chat room (although some can be set up to do so). If you learn that your child has installed an IRC chat client on a computer, you should ask why he or she has installed the software and how it's being used. What groups are being visited (they may be saved as favorites in the client software)? What kinds of conversations are occurring? Is any other activity taking place (file uploads or downloads, streaming video, etc.)? Has your child met or made plans to meet someone from the chat group?

There's no question that monitoring your child's communication on the Web is intrinsically more difficult than monitoring other forms of communication. The biggest challenge is simply keeping track of where that communication is taking place. Without monitoring software, it can be difficult to know if your child has posted something objectionable in a Google message group, or has uploaded a salacious photo to one or more of the dozens of media-sharing sites, social networks, and teen communities that spring up like weeds.

Of course if your child is using a computer for more serious types of crimes—hacking, identity theft, computer fraud, and the like—he or she may be going to great lengths to hide the activity. If your child is spending a great deal of time online and you have no idea what's going on in his or her virtual world, or if your child is younger and simply needs more guidance on appropriate social behavior, it might be appropriate to install active monitoring software on the computer. In chapter 16, there's a discussion of third-party software tools that can capture that information and send updates to your inbox.

Ultimately, when parents worry about knowing what their child is doing online, what they're really asking themselves is how well they know their child. That's a question best answered one electronics-free dinner at a time

Chapter Three
There Is Such a Thing as Too Much Sharing

One of the earliest lessons we teach our children is the importance of sharing. It starts with siblings, quickly progresses to neighborhood playmates, and then preschool and kindergarten classmates. The values inherent in the act of sharing—generosity, kindness, empathy, community— are crucial to a child's development, and important to society as a whole.

Sharing, however, is not an unalloyed virtue. As children get older, it becomes necessary to teach them some of the nuances of sharing: don't share the soccer ball with members of the opposing team; don't share more than you can afford; don't share things that aren't yours to begin with; don't let others take advantage of your generosity. The concept of sharing quickly gets complicated.

When I was a kid in the 1970s, parents—mine included— were feeling a lot of anxiety about their kids sharing too much personal information and falling victim to sexual predators the news media insisted were lurking behind every tree, even in relatively quiet suburban towns like the one in which I grew up. I was warned by my parents never to give out our address, nor our home phone number. I was even supposed to keep mum about the species and names of our pets, though I'm not sure what my parents thought the sexual predators, mostly mythical, might do with such information.

Back then, there were no easily forwardable e-mails nor text messages, no digital technology that could be used to rapidly distribute compromising and suggestive photographs. Our "social networks" didn't extend much beyond our neighborhood. Nevertheless, we were raised with the civilian equivalent of the military motto, "loose lips sink ships."

A Vast Web of Personal Information

A good case can be made that the news media still overhypes the threat of sexual predators—sex sells, after all, and sex crimes sell even better—but the examples of kids getting into trouble online are indeed real, and far too numerous. There are plenty of reasons you might not want your children sharing too much of their personal information online, whether your preferred boogeyman is an updated version of that 1970s-era sexual predator (now with Internet access) or an overly aggressive corporation.

It's unrealistic, of course, to keep our children off the Internet altogether, and it's counterproductive to raise them in a fog of fear, but it's a valuable exercise to make them think about what information they're sharing, who has access to it, and what the consequences could be if it reaches a larger-than-intended audience. Our hyper-connected world has made it increasingly difficult for children to follow that ages-old lesson: don't talk to strangers. It's a near certainty, in fact, that strangers know all sorts of things about our children, though they may not be using that information for any more nefarious purposes than selling them the latest clothing fad or video game.

Social networking sites like Facebook are a perfect example of how much personal information your child could be sharing online, both intentionally and unintentionally. When someone first joins Facebook, for instance, the site asks for a rather staggering amount of information, all of which it will then share with the user's friends and networks:

- sex
- birth date
- children, siblings, and other relatives (with links if they, too, are on Facebook)
- relationship status
- current city
- hometown
- education and work history
- activities
- interests
- favorite music, books, and movies
- contact information (home and mobile-phone numbers, address, e-mail addresses, personal websites)

And that's just the basics. The site encourages users to upload photographs and to tell others, via status updates, where they are and what they're doing. That's to say little of the near-infinite number of Facebook modules and third-party applications that relentlessly query for information about a user's likes and dislikes. In the wrong hands, or in a misguided moment, all this information could become fodder for harassment and abuse.

In fairness, Facebook is merely the scapegoat for a number of similar Internet venues for information collection and inappropriate disclosure. There are hundreds, if not thousands, of smaller websites that routinely collect and analyze data about teens and preteens. Exactly how much information is collected and just how far it spreads varies from site to site. As parents, our first challenge is knowing which sites our children have joined, and what information they've disclosed.

In theory, at least, there are legal protections for the online activity of children under the age of thirteen. As of April 21, 2000, any website that specifically markets to children under thirteen, or that knows young children regularly provide it with information, must comply with the terms of the Children's Online Privacy Protection Act of 1998 (COPPA). Among other provisions, COPPA requires such websites to have specific privacy policies in place, to obtain consent (with certain exceptions) from a parent or guardian, and to take steps to protect child privacy and safety. The Federal Trade Commission offers guidance on the application of COPPA on its website (http://www.ftc.gov/privacy/coppafaqs.shtm).

Websites can run afoul of COPPA by not properly handling a child's personal information, and the FTC has handed down substantial fines to companies for doing so. In 2006, for instance, the social networking site Xanga agreed to pay a $1 million civil penalty for "collecting, using, and disclosing information from children under the age of 13 without first notifying parents and obtaining their consent." Two years earlier, UMG Recordings paid a $400,000 fine for similar violations on a large number of music-related sites.

Unfortunately, there are far too many websites for the government to effectively police them all, which means the responsibility for knowing what children are sharing online lies first and foremost with parents. There are software tools that can help, but the best tool continues to be frank and frequent discussions with our children about what exactly they're doing online and with what electronic devices.

Children may be knowingly sharing personal information online—though often without thinking through the potential consequences—but an increasing amount of this information is being collected online without the knowledge of either children or their parents. For instance, a few years ago, Facebook introduced its iconic "Like" button, which gave users the ability to show their approval of friends' posts, photographs, causes, and so on. The company was so enthusiastic about its little blue "Like" button that it offered the technology to other websites. Facebook users can now "Like" just about any sort of content imaginable, from news articles to blog posts to photo albums. When you find something you like on, say, the *New York Times* website, or the *Huffington Post,* a simple click of the "Like" button will prompt Facebook to update your profile so your friends can follow the progress of your mid-morning procrastinations. It should come as no great shock that Facebook is not only carefully collecting and analyzing data about what its users like, regardless of where on the Web their votes are cast, but it is also making it available to advertising networks like Google's DoubleClick (already one of the largest online ad networks). This information is electronic gold for online advertisers, who view it as the most reliable data available about what you—or your children—are interested in at any given moment.

As a recent *Wall Street Journal* article pointed out, mobile apps are particularly aggressive in their collection of personal and location-based data, and these apps often neglect to inform users that such data is being collected. Depending on the service in question, location-based information can be combined with other mobile activity to create a startlingly detailed profile of a given user. For instance, the *Journal* reported that the mobile version of Pandora "sent age, gender, location, and phone identifiers to various ad networks." Similarly, textPlus 4—an iPhone texting app that gives texting capability to WiFi-only devices like the iPod Touch—distributes the unique ID of each device and the user's area code to eight different advertising networks. Advertisers are willing to pay a premium for geo-location data, on the theory that the most effective ad is one that reaches potential consumers when they're close to the business doing the advertising.

The good news about this kind of geo-location data—as creepy as it might seem—is that it's largely transactional in nature, meaning there's little chance that someone other than an advertiser or social network administrator could get hold of it.

However, there are a growing number of "check-in" services—Google Latitude, Foursquare, Yelp, and Facebook Places, for instance—that encourage users to make their locations available to their entire social network in real time. During the course of a single day, a teen or preteen might "check in" at the mall, at a coffee shop, and then at the local fast food joint (Hansel and Gretel should have been so lucky). Many businesses, in fact, are offering customers coupons and special deals if they "check in" when they visit.

There are other ways your child could be disclosing information—often unintentionally—which could in turn be misused by friends or classmates. Virtually all photos taken by smart phones, for instance, and by many recent-model cameras, are embedded with information known as metadata that identifies the date and time the photo was taken, as well as the location. That information can easily be read by websites like Google Maps, Facebook, and Flickr, which use it to populate their geo-location features.

To see a typical example of how these features work, just go to Google Maps, enter one of your favorite destinations, and click on the "Photos" button on the right side of the map. If it's a popular destination, you'll see dozens of photo icons pop up on the map. Click any one of them and you'll be transferred to a page on Google's photo-sharing site, Panoramio. There, you can view the exact longitude and latitude of the location where the photo was taken, as well as a link to the profile of the person who took it.

It's not hard to imagine how useful all this information could be to a potential bully. Granted, most bullying incidents occur among school classmates—and it's not that difficult for classmates to find one another—but the availability of geo-location information and the popularity of "check-in" services like Foursquare, Yelp, and Facebook Places add yet another tool for harassment.

Few parents want to contemplate the possibility, but there's another use of geo-location information that can affect their children: that is, as part of a criminal investigation. Mobile-phone tracking data has become a powerful tool for police, who can use it not only to identify where a particular mobile device was at a given date and time, but also to track its movements over a period of time. If a perpetrator is foolish enough to post mobile photos to a social networking site (as many graffiti taggers have been known to do), those photos' metadata will act as an evidentiary lock and key.

Misuse of Others' Personal Information: A Serious Cybertrap

There are many reasons, obviously, to be concerned about your child sharing too much personal information online. The more serious legal issues, however, are those surrounding what your child does with the personal information of others. There can be serious legal consequences when a child violates someone else's privacy rights.

It's difficult for parents to teach their children to respect others' privacy in the offline world—explaining, for instance, why they should resist the urge to read a sibling's diary, or why it's wrong to spread secrets about a classmate. That job becomes even more difficult the minute our children go online. One problem is the Internet itself, the very structure of which seems designed to promote a cavalier attitude toward the ownership of information, whether it's the latest hot song or a particularly juicy piece of gossip.

1. Civil Liability for Misuse of Shared Information

Many people don't realize just how much trouble kids can get into by misusing information other kids have put online. The most common scenario is the one discussed above, in which a child uses someone else's online information for the purposes of harassment or bullying. Bullies might dig up information their victims didn't know was being collected, such as geo-location data, but far more often bullies simply make use of information the victim has purposely posted online. A person's likes and dislikes, choice of friends, or sense of style, can all be used to make a victim feel excluded and unpopular. In that way, the Internet—and the phenomenon of social networking in particular—has created what the military might refer to as a "target-rich environment."

As we'll discuss in more detail in chapter 6, cyberharassment and cyberbullying are typically prohibited by school policy and, increasingly, by state law. However, those are not the only laws that govern the misuse of someone's personal information. Another source of potential liability occurs when a child shares private, personal information about someone else in a public forum, most typically on a social networking site like Facebook or MySpace. The most common scenario is dissemination of gossip not intended for general distribution, but liability could also result from the redistribution of private photos (i.e., any unauthorized photos, up to and including those with nudity or sexual activity [better known as 'sexting']).

Depending on the circumstances of a particular case, the damages caused by an unauthorized disclosure of information can expose a child and his family to significant civil liability and possibly even criminal penalties. Some states allow people to sue for civil damages for invasion of privacy. Other states permit lawsuits to be filed against an individual who depicts another in a false light, although that claim usually requires proof by the plaintiff that the defendant acted with actual malice. In extreme cases, a victim or his family can even sue for intentional infliction of emotional distress. Those suits are growing increasingly common, particularly in cases involving significant harm or death to the victim, such as that of Tyler Clementi, a former Rutgers University student who took his life following the online posting of a hurtful video (discussed below).

According to Internet legal specialist Parry Aftab, who runs the website WiredSafety.org, approximately one hundred such lawsuits were filed in the twelve months prior to October 2009. Unfortunately, in the fourteen months since, there have been a number of high-profile teen suicides resulting from online harassment, which makes it even more likely that victims and their families will look seriously to legal remedies.

2. Criminal Conduct

There are circumstances in which sharing someone else's secret or personal information can be construed as a criminal, not just a civil, violation. Whether a prosecutor chooses to file criminal charges generally depends on the seriousness of the injury and the amount of attention the case receives in the news media.

As mentioned above, a particularly sad and painful example of unauthorized information disclosure occurred in the fall of 2010 at Rutgers University, when two freshmen—Dharun Ravi and Molly Wei— allegedly used a webcam to surreptitiously film and broadcast a video of Ravi's roommate, Tyler Clementi, engaged in a sexual encounter with another male student. A short time later, Clementi updated his Facebook status to read "Jumping off the gw bridge sorry," and then proceeded to do so.

Following Clementi's death, Ravi and Wei withdrew from Rutgers, and New Jersey prosecutors charged the two students with third-degree violations of the state's privacy law, which makes it a crime to surreptitiously view—and record—nudity or sexual contact. Given the facts of the Rutgers case, there is some question as to whether the alleged crime can be proven beyond a reasonable doubt, but even if the prosecutors fall short in the criminal trial, there is little doubt that Ravi and Wei will face substantial civil claims for intentional infliction of emotional distress. In any case, none of this adequately contemplates the tragedy of the talented young violinist's death.

A more common scenario involving the misuse of someone else's private information is the typical sexting case (a crime discussed in more detail in chapter ten). Sexting refers to the sending of nude or semi-nude photos from one mobile device to another, and if it occurs between minors—even consenting minors—it is a criminal act. It becomes an even more serious crime when the relationship founders and one person forwards the explicit photos to a larger audience.

At least a half-dozen suicides have been attributed to sexting, the most infamous being the 2009 death of Ohio teenager Jesse Logan, who hanged herself after suffering months of harassment and abuse from classmates who saw a nude photo of her distributed electronically by her ex-boyfriend. The publicity surrounding her case and others has led many state prosecutors to bring criminal charges not only against the children who actually distribute the photos, but also sometimes the children who took the photos in the first place, even if they had no intention of distributing them.

Neither the Clementi nor the Logan case may be appropriate to share with young children just beginning to explore the online world. But the minute your children are entrusted with a camera phone or Internet access, it becomes vital to teach them the importance of discretion and the art of keeping confidences—not those, obviously, that hide a crime, or could result in someone being hurt, but certainly those that are matters of privacy. It's equally important for you to spend enough time online—and, in particular, on social networking sites—to appreciate just how easy it is for your kids to get access to the private information of others and share it. That will educate you about the risks your children face, number one, and it also may give you second thoughts about how much of your own information you want to post online. The truth is we could all probably do with a little more discretion.

Chapter Four
Who's in Your Child's Sharing Circle?

Knowing what information our children are sharing online is only half the battle; the other critical component is having some idea of with whom they're sharing that information, either intentionally or accidentally. To a large degree, it is the scope of a child's electronic sharing circle (and who's in it) that will determine what legal liability he or she might face for releasing problematic content online.

A cruel, hurtful message sent via text from one child to another almost certainly constitutes harassment or cyberbullying, but if the same message is broadcast to a large number of people on a blog or social networking site, it could also be construed as libel or defamation. Similarly, a court might conclude that a bullying victim's damages for pain and suffering should rise in proportion to the number of people who read his cruel and harmful post.

It's important to appreciate just how quickly and easily children can build surprisingly large audiences for their messages. Given the viral nature of social networking sites, it's not uncommon for kids to have several hundred "friends" on Facebook, and they can engage equally large (or larger) followings on microblogging services like Twitter, or by starting and maintaining their own more substantive blogs.

The ability of kids these days to quickly and easily reach large audiences represents a fundamental change from even a few years ago (Facebook, after all, wasn't even created until 2004) and a complete sea change from a few decades ago. When I was growing up in the 1970s, the only way to communicate with more than a few people at a time was to stick a typed or handwritten message to a school bulletin board, hang fliers around town, spray-paint a wall, or start an underground newspaper (the last of which, actually, some friends and I did, though the only targets of our sophomoric humor were a handful of teachers and administrators). Those antiquated types of communication could lead to many of the same legal problems I'll discuss in the next section of the book, but there are several key differences between the publication options I had thirty years ago and the ones kids have today: incredible speed, low cost, staggering reach, and in general, far less supervision.

Thirty years ago, reaching a large audience required at least some investment of time and resources. The cheapest options were bulletin-board posts (the physical type that require a pushpin) or a can of spray paint emptied onto a blank wall. However, these were subject to rapid official intervention and, in the case of graffiti, serious retribution if one was caught. Fliers and underground newspapers had the advantage of generally being lawful expressions of free speech, but printing costs were fairly expensive.

Kids today have enormously powerful resources at their fingertips, resources which can be used to send information quickly and cheaply, allowing them to reach an audience of thousands or even millions within seconds. This greatly amplifies the potential damage kids can cause—and the legal consequences they can face—through a malicious or recklessly hurtful message.

Given the fluid and ever-changing online relationships your child undoubtedly maintains, it's simply not realistic to expect—as my parents did—that you will know everyone your child considers a "friend" on Facebook or other social networking sites. You certainly won't know the identity of everyone who reads your child's blog posts or tweets. In order to effectively assess the cybertraps endangering our children, it's important that we have at least working knowledge of who they communicate with most frequently, and through which medium.

Texting Contacts

Given the rise of texting as the preferred form of communication among teens and preteens, your child's most significant audience is likely to consist of the friends and acquaintances with whom he or she regularly exchanges these short, frequent missives. Since the overwhelming majority of texts are sent from one mobile phone to another (they can be sent from a phone to an e-mail account and vice-versa, but few teens actually use this option), a kid's text buddies typically will be a subset of the contact list saved on his or her phone.

Every mobile phone has the capability to display text messages it has sent and received, as well as the identity of the people who've sent and received them. In an older (non-smart) phone, the messages may simply be shown in chronological order. Newer phones, like the iPhone and Android, organize text messages by an individual's name, so that one click on a name will show you the entire text history with that person. My iPhone, for instance, makes it possible for me to quickly scroll through months of text conversations I've had with my sons.

It's not a bad idea to periodically review your child's mobile phone, to get some idea of who's on the phone's contact list, what kinds of text conversations are occurring, and what messaging apps have been installed (if it's a smartphone). While privacy is obviously an important value to teach children, and something to which they are entitled—to varying degrees—it's important to remember that privacy is a privilege, rather than a right, at least until a child turns eighteen. At least as important as our children's privacy, if not more so, is protecting them from the various cybertraps they may not fully understand or appreciate.

If your child is in the habit of deleting his or her text messages, a simple examination of the phone won't reveal frequent text contacts. (But since most kids don't take the time to routinely scrub their message history, an empty history is a bit of a red flag.) Depending on your relationship with your child, you may be able to simply ask for a list of names. If that's not possible, or if you have concerns that the list is not as comprehensive as it should be, the simplest method of identifying your child's text buddies is to get a copy of your mobile bill, either on paper or online, and look for the text message information linked to your child's phone number. AT&T Inc. labels that information as "Data Detail," and even lists the phone numbers of each incoming and outgoing text.

One immediate problem, however, is that a "Data Detail" only contains phone numbers, not names, so unless you have the numbers of your child's friends memorized it may not be possible to identify the most frequent texters right away. You do have a couple options. If you have access to your child's phone, you can compare phone numbers in the contact list with the ones listed on the phone bill. Or, if you've requested copies of the actual texts from the phone company, the identity of your child's correspondents may be clear from the messages themselves. Another option is to sign up for a membership with one of the various reverse mobile-phone look-up services, and simply start plugging in numbers. At the very least, you'll get the name of the person (usually a parent or guardian) paying for the phone.

How necessary is this kind of effort, time, and expense? The answer depends on the relationship you have with your child and the concerns you have about his or her peer interactions. If you recognize that some conflict is taking place, or your child is noticeably upset, text messages may well be a factor—either because kids are gossiping about the conflict via text, or because text messages themselves are the weapon of choice for bullying or harassment. In either case, it is irresponsible not to know who the players are in your child's mobile world.

Social Network Friends and Followers

On the basis of numbers alone, keeping track of your child's social networking friends is not easy. Typically, kids will accept "friend" requests from virtually everyone in their middle or high school, so your child's Facebook "network" can easily reach several hundred users. Luckily, there's a limit to the number of people even the most Facebook-obsessed kid can interact with, so your child's core group of online friends is likely much smaller. What's less encouraging is that unless your child has spent a lot of time fine tuning his or her privacy settings, all those "friends" can see everything he or she posts.

There's a lot of value to sitting down with your child and going through the privacy options for Facebook or other social network sites on which he or she spends time. Online companies have slowly gotten more sensitive to consumer concerns about privacy, and although the options aren't perfect, they're at least easier to find and manipulate than they used to be. For instance, Facebook has a "Privacy Options" link in the pull-down menu for the Account of each person (currently in the upper-right corner of your Facebook page). For each category of information you post to Facebook—status updates and photos, bio and favorite quotations, family and relationships, religion, etc.—you can grant access in descending order to "everyone," "friends of friends," or just "friends."

Given the global reach of social networking sites and the potential for content to migrate to the rest of the Internet, it's particularly important to be aware of your child's online friends and what information they can access. One of the most important tools for keeping an eye on social network activity is to establish and enforce a mandatory friend policy.

The basic concept is simple enough: if your teens (or pre-teens) sign up for any online service or social networking site, the first thing they need to do is send you an invitation to join them. For as long as they remain members of the service or site, they must grant you the fullest possible access to the information they post. It would be useful, I think, if social networking sites implemented a "parent" or "guardian" status for accounts, but that is probably a bigger headache than the sites presently wish to deal with.

Requiring full friend status on your child's social networks will only work if you're confident about which social networking sites your children have joined. Ideally, they will honor an agreement to tell you as one condition of using a computer or mobile device in the first place. But if you have doubts—or fear your child is being secretive about his or her online activity—it may be necessary to employ visual inspections or even electronic surveillance.

Let's assume for the moment that your child is upfront with you about the sites they've joined, and willing to grant you "friend" status on those sites. What will you be able to glean from their online activity? When my sons both joined Facebook a few years ago, I enforced the "friend" rule, which means I'm able to view their status updates, wall postings, and photos. That access has given me useful insights into issues they're facing, things they might not discuss over dinner, and occasionally I've had the chance to point out things they've said that could be hurtful or reflect poorly on them. It's been useful, for instance, to encourage my younger son to clean up his online language by reminding him that not just his dad, but his aunts and grandmother, are among his Facebook friends. There was one brief period when my older son tinkered with his privacy settings and "forgot" to include me in his inner circle, but when asked about it, he quickly restored the default. There's no question that kids (including my own) express annoyance about this level of supervision, but, in the long run, I believe it comforts them to know that someone is looking out for them and cares about what they're doing online

Unlike text messaging, social networking sites rely entirely on names, so if you have full friend status with your child, it's easy to see the identity of all his or her friends on that particular social networking site. (Incidentally, full friend status doesn't have to be reciprocal; with a little bit of tweaking, you can limit what your children see on *your* Facebook page.) Maintaining full friend status with your children on social networks allows you to quickly and easily get a sense of their public interactions with their peers. Who are their close friends? Who might they be having conflicts with? What types of issues are they discussing or wrestling with? Have you seen something that you think another parent should be aware of?

It's worth noting that Facebook does have both a public component (status updates and wall posts, which everyone can see) and a private component (messages, which go directly from one Facebook user to another, much like e-mail). If you're logged into your own Facebook account, you can see only your child's public interactions. In order to see private messages, you would need to log into your child's account, which would require an ID and password.

Obviously, asking your child for a password is a more serious—and potentially more confrontational—privacy intrusion than simply requiring full friend status. Whether you think it's necessary to have that level of access to your child's social networking memberships will depend entirely on your assessment of what is happening in your child's life, and your concern about any possible harm he or she may be facing, or inflicting on others.

Instant Messaging, Chat Room, and Video Chat Buddies

Here instant messaging, chat rooms, and video chat services are lumped together for one main reason: each is a method of conducting a real-time conversation with one or more individuals. Traditionally, these types of conversations have been PC-based, but it's increasingly possible for children to engage in these activities on their mobile devices as well.

Your child's contact list for instant messaging is likely to overlap with his or her other contact lists. For instance, online e-mail services like AOL and Google have built-in instant messaging functions, so logging in to an e-mail account will turn up a list of contacts who are online and available for chat. The same is true for most social networking sites (e.g., Facebook or MySpace). Not everyone sets up their e-mail account or social networking page to enable instant messaging, but it's safe to assume your child can instant message with anyone who is on the same e-mail or social networking service.

Chat rooms are particularly dangerous. If your child enjoys visiting chat rooms, it will be extremely difficult to know the identity of the people with whom he or she is chatting. Virtually everyone who visits a chat room uses some type of handle or ID that's meant to mask his or her real identity. More than that, the software your child uses to connect to a chat room—usually a browser, but sometimes stand-alone software—does not bother keeping a list of chat contacts. Since chat room users do not "subscribe" to a particular chat room, but instead just log in and out when they want to chat, there's no real point in creating and maintaining a contact list.

The fact that identities are so difficult to determine in chat rooms underscores the importance of knowing what your child is saying when he or she visits them. If it's relatively harmless chatter about how to move to the next level of World of Warcraft or a discussion of the errors and inconsistencies in the *Lord of the Rings* movies, that's one thing. But if your child is asking how to order smokeable herbs online or disclosing to complete strangers his or her "a/s/l"—age, sex, location—then the lack of identifying information may make it difficult to defend your children against those who may be exposing them to harm.

Video chat may be even more of a risk. Most of the major chat tools now allow users to broadcast images from a webcam while chatting (both Google and AOL have this capability), and there are standalone programs like Skype specifically designed for video chat. Much like text-based instant messaging, you can generally assume that the people with whom your children video-chat are in their email contact lists or, in the case of Skype, in a separate list of contacts maintained by that program.

Meanwhile, in a development anticipated years ago by the comic detective Dick Tracy, kids who have the latest-model iPhone 4 can talk to other iPhone 4 users with "FaceTime," a feature that enables callers to see video of each other while they talk. If your child has access to the technology, you won't be able to see a recording of what was said or shown, but you will at least be able to see who took part in the conversation, either in your child's recent call history or on the phone bill.

There are also several video chat rooms your child can visit. Some of the most popular—Stickam, Camfrog, Chatablanca, Meebo, and Tinychat—allow both one-to-one video chatting and one-to-many (where multiple users log in to watch one user's broadcast). In most cases, the video chat sites are designed as communities, so you can see a list of the people your child has friended. But keep in mind that, as with text-based chat rooms, many people drop in and out of these video chat forums without ever identifying themselves. Your child could easily be having long and explicit conversations with someone halfway around the world, or just three blocks over. Remember: for services that do full video chat—both video and audio, rather than just video and typing—there will be no record of what was said. The only information you'll have is that your child visited a particular chat site.

One particular video-chat site that's garnered a lot of recent attention is Chatroulette.com, a service that connects video chatters to each other at random. If you log in to the site, you'll immediately be shown another, random user's video feed. If that person looks like someone you want to chat with, the two of you can type messages back and forth. If you're not interested, or you grow bored, you can simply click the "Next" button to be whisked away to yet another random feed. As various news outlets have noted, a not-insignificant number of the video feeds feature undressed men fondling themselves. Fortunately, this particular service doesn't appear to have captured the teen imagination, more because it's too unfocused rather than because of any hesitancy about showing explicit content online. Still, while Chatroulette can be viewed as an interesting technological and social experiment, there's probably not much good that can come from your child spending time there.

Twitter Followers and Blog Subscribers

The final major category of people with whom your children may regularly interact online is made up of those who follow their Twitter feeds or read their blog posts. As with chat rooms, there's likely no reliable way to know the identity of all the people following a given blog or feed, so the focus here should be on what's being broadcast by your child, what information he or she is revealing to the world, and what's being said about others. For instance, both Twitter and traditional blogs can be used to make statements that might constitute cyberbullying, cyberharassment, or defamation (for a detailed discussion, please see chapter seven).

It's not impossible for anonymous conversations to arise out of Twitter or blog posts, but there are a few built-in protections. For instance, someone following your child's Twitter account cannot send a direct message unless your child is also following that person. Similarly, comments can be blocked on a blog, preventing random trolls from posting harassing messages. By and large, someone who wants to respond to your child's blogging or tweeting online will have to communicate in a manner that you'll be able to monitor.

Part Two: The Cybertraps

Chapter Five
Intellectual Dishonesty and Intellectual Property Theft

In the "offline" world, the lesson that stealing is wrong begins early. Before our children start preschool we teach them not to take toys from each other in the living room or at daycare. As they get older, they're taught not to go fishing around in each other's rooms, lockers, or backpacks for items that don't belong to them. Most kids, by the time they're old enough to walk around a mall without a chaperone, are clear on the concept that it's illegal to go into a music store and come out with a stack of CDs hidden under a coat.

The concept of intellectual honesty is slightly more abstract, but by the time kids are in middle school relatively few would do something so blatant as copy pages from a book and try to pass them off as their own work—or, if they did, most would at least understand that they were doing something wrong, and that if they were caught there would be consequences.

Not only are these real-world examples of honesty easy to understand, but violations tend to be dealt with quickly. If a child goes into her sister's room and "borrows" a favorite sweater, the protest will be swift and loud. Teachers, for their part, have long experience recognizing the often-stilted language of encyclopedias and textbooks. And theft-prevention devices employed by music and department stores generally do a good job identifying shoplifters.

In cyberspace, however, the gap between crime and punishment can seem infinite. Given the millions of potential sources online, just how likely is it that a teacher will stumble upon the one website from which a student lifted material? How likely is it that a record label will pursue *your* child for illegally downloading music? How will Adobe even notice if one more illegal copy of Photoshop or Dreamweaver finds its way onto a laptop?

These cybercrimes, however, are just as problematic as their real-world counterparts, and like their real-world counterparts they can carry serious penalties. Your child might feel as if the odds of getting caught are miniscule, but the means of policing online behavior are growing more sophisticated all the time. Ironically, copyright holders typically use the same technology to catch pirates that children use to commit Internet theft in the first place.

Plagiarism
1. The Offense
Plagiarism, in a nutshell, is any attempt to pass off someone else's words or ideas as your own—to take advantage of another's work without crediting the source. Plagiarism is not generally treated as a criminal offense, but rather as an ethical violation and a breach of school policy. However, it *is* a form of copyright infringement and, in extreme cases, plagiarism could be prosecuted under either federal or state law.

Most middle- and high-school students agree to abide by an academic honesty policy that defines plagiarism and lists the consequences for intellectual dishonesty (you can check with your school administrators for a copy of the policy or policies that apply to your child). Teachers also generally discuss plagiarism with their students, particularly once those students begin to take on more research-intensive projects. There are gray areas, of course. Teachers are often understanding about unintentional plagiarism—a student trying to put something into his or her own words but sticking a little too closely to the original, or forgetting to cite a specific fact. For intentional plagiarism, however—the wholesale copying of another's work—the punishment can be quite serious.

2. The Scope of the Problem

It's hard to know exactly how much plagiarism occurs in the nation's middle and high schools, and any estimates are complicated by the fact that plagiarism isn't always easy to identify. As American playwright Wilson Mizner once said, "If you copy from one person, it's plagiarism; if you copy from two, it's research."

The open nature of the Internet and the vast amount of information online (much of it published anonymously) simply confounds the problem of plagiarism, since it encourages the belief that all that information is free and available for the taking, that no one really "owns" it, so appropriating it for our own devices is just how information is exchanged in an electronic age. Students often commit minor acts of plagiarism without even realizing it, and this is an area where teacher and parent intervention can help.

The larger issue, however, is intentional plagiarism, where a student knows that he or she has crossed the line from research to literary theft. Sure, there are gray areas, but most students know they're not allowed, for instance, to simply copy and paste text from the Internet into their term papers. Then there are the many websites that offer to sell students papers written by others—for "research purposes," of course. Savvy students can also find papers using simple Google searches (you'd be amazed how many high school term papers wind up online).

In a study conducted recently by University of Rutgers business professor Donald L. McCabe, nearly half of all high school students admitted they have plagiarized material from the Internet. Even more sobering: of those, nearly one-third said they didn't consider copying web content to be cheating.

3. Investigative Tools

Ironically, the same technology that makes it so easy to commit plagiarism is making it easier to detect. I'm not talking about those obvious instances in which a student copies content from a website and forgets to match up the font type or color with the rest of the paper (when a paper is typed in black ink and there's a big section in purple, it's a bit of a give-away). Instead I'm referring to the growing number of online tools teachers are using in their efforts to enforce academic honesty.

The obvious starting place is a search engine like Google, Yahoo, or Bing. It's a pretty simple measure to take a sentence or two from a student's paper, type it into the search engine in quotes, and see if a match appears. Some teachers will test papers at random, but it's more likely they'll look closely at papers when they detect a sudden change in tone or style: passages that don't match the student's own voice. The longer a teacher works with a particular student, the easier it will be to recognize a piece of that student's writing that seems out of character.

Teachers and administrators also can use the same sites that make it easy for your children to purchase papers. Students, after all, aren't the only ones with instant access to ready-made term papers, and if a teacher suspects cheating it takes only a few minutes to search the more popular sites—PerfectTermPapers.com, LotsofEssays.com, or TermPapersCorner.com—for a match.

A more comprehensive hi-tech solution is offered by the website Turnitin.com, which is quickly becoming the industry leader in plagiarism detection. The fifteen-year-old company offers a set of software tools to academic organizations that includes a paper upload module that cross-checks a given paper against the company's large—and growing—database of content. A teacher can upload a paper to the site and within minutes see a report detailing any "borrowed" language in the student's work, whether that language was lifted from a website or printed article. Turnitin will even check papers against the growing cache of student assignments uploaded to its site by teachers around the world. The company says it already has a database of 135 million student papers, as well as ninety thousand journals, scholarly papers, and books, not to mention an index of nearly fourteen billion web pages.

Turnitin's practice of storing and cross-referencing student papers actually sparked a 2007 copyright infringement lawsuit by four high-school students, but the suit was thrown out by a U.S. District Court in Virginia in 2008. According to the court's decision, Turnitin's use of the student papers constitutes "fair use" and provides a "substantial public benefit" by discouraging plagiarism.

4. Consequences

Although plagiarism—at least to a child—may seem like a "victimless crime," rest assured that educators take it quite seriously, and the consequences for academic dishonesty become more severe as your child progresses in his or her academic career. At the very least, a student will typically receive zero credit for a plagiarized assignment, and in more blatant cases (or repeat offenses) a student can face suspension or even expulsion. Students who have been convicted of plagiarism or other acts of intellectual dishonesty will find it's more difficult to get into a quality college or university, since such offenses will often be noted in the student's academic record.

Schools take a wide range of approaches when it comes to plagiarism. You can see some of those approaches and sample academic honesty policies on the book's website at www.cybertrapsfortheyoung.com.

5. Prevention

Given how easy it is to copy sentences and paragraphs from a website to a student paper, is there any reliable method parents can use to stop their children from plagiarizing? There are keylogging tools that can help, but in this particular instance the most effective solutions are the old-fashioned ones: a solid moral education and a good example.

Since this book focuses on the crimes children can commit online, it may come as little surprise that morality is an underlying theme. It's somewhat easier to make a moral argument when the activity being discussed—cyberbullying, hacking, phishing, sexting—is a crime with clear, concrete consequences. The chief difficulty with plagiarism is that, to kids, the consequences can seem both minimal and abstract—particularly in the short term—compared to the perceived benefits: exerting less effort on a given assignment and receiving better grades.

It doesn't help, of course, that the Internet has cultivated such a widespread culture of copying. Most kids don't get to see their parents in a classroom setting, but they do pay close attention to the things their parents do at home, including their online activity. How effectively can you discuss the issue of plagiarism with your kids when they know (and trust me, they do) that two-thirds of your music collection was illegally downloaded, or that last night the whole family snuggled up to watch a movie Dad pirated from a torrent site? It certainly wasn't middle- or high-school kids who crushed the music industry—it was mostly twenty- and thirty-something adults, many of whom are now parents themselves.

Academic honesty begins at home, both in conversation and in role modeling. Kids are often under the impression that grades are the only thing that matters—to their parents, to colleges, or even to society as a whole—and that kind of pressure for results can certainly motivate children to cheat. It's important to talk with your kids about the real benefits inherent in education—the skills and knowledge that come from dedicated effort—and clarify that those benefits are only enjoyed if your kids are doing their own work.

Copyright Infringement

One illegal download at a time, the culture of copying spawned by the Internet has nearly destroyed the music industry. Television networks, meanwhile, are racing to adapt to the new entertainment paradigm, and the book industry is rapidly discovering that the popularity of e-books, while lucrative, will result in exposure to many of the same risks faced by their audio and video peers.

Even more so than plagiarism, copyright infringement can seem like a victimless crime. The rationalizations are numerous: it's just a single download; Lada Gaga is already a multi-millionaire; Sony makes billions of dollars every year; everyone does it. Most of these statements are true, of course, but they don't alter the fact that copyright infringement is prohibited by both federal and state law.

1. The Offense

According to the United States Copyright Office, "copyright infringement occurs when a copyrighted work is reproduced, distributed, performed, publicly displayed, or made into a derivative work without the permission of the copyright owner." Copyright law in the United States is governed by the provisions of Title 17 of the U.S. Code; extensive information about copyrights can be found on the web site of the U.S. Copyright Office at www.copyright.gov.

Copyright infringement occurs in myriad ways, but these days the most common is the downloading of audio and movie files through peer-to-peer networks, torrent sites, or online file repositories like RapidShare or MegaUpload. In the case of peer-to-peer software, there is the added complication that the software is typically set up to share downloaded files by default, which increases the potential penalties for copyright infringement.

2. The Scope of the Problem

Back in 1999, Shawn Fanning created a program called Napster, which enabled people to more efficiently share music online. The music industry—represented chiefly by the Recording Industry Association of America and the various major music labels—filed suit against Napster, alleging that it was infringing on their copyrights by storing and distributing protected content. The federal courts agreed, and Napster was shut down just over a year after it was launched. But the damage was largely done, and while the RIAA was able to rid the world of Napster, the next generation of file-sharing services has proven more difficult to stop.

Not long after Napster was shut down, various programmers developed peer-to-peer (P2P) networks that could be used to share files without the need for central servers. By using software programs known as P2P clients (the now-defunct LimeWire and FrostWire are two examples), music lovers could search for and share music files directly from each other's computers. Instead of a central repository of files stored on a single, massive server, P2P networks create a giant web of interconnected computers—a microcosm of the Internet itself—where users can swap files with each other, directly from one hard drive to another. The software doesn't actually hold any of the files; it only creates the connections.

The development of these P2P networks has made copyright enforcement much more difficult for the music and film industries, since there is no single, central server that can be shut down. While this might seem like good news for illegal downloaders, there's bad news, too: rather than giving up the fight, copyright owners like the RIAA are now choosing to go after individual P2P users who are illegally downloading and sharing content. Over the past decade, tens of thousands of lawsuits have been filed against individuals (including many children) who have been detected either downloading or sharing protected content online.

It is, admittedly, an uphill battle for the music industry. Before being shut down by a federal judge last month, the popular P2P software LimeWire had roughly fifty million users, and the number of pirated songs, movies, and software estimated to be circulating on P2P networks is in the billions. There are conflicting reports on the actual financial impact this has had on copyright holders, but the common consensus seems to be that online piracy has cost the music industry—record companies, producers, and artists—somewhere between $6 billion and $10 billion in revenues over the last ten years. That calculation is complicated by other factors, including falling CD prices and the shift from albums to the sale of individual songs, but it makes intuitive sense that if people are downloading music for free, most are not paying for a second copy.

3. Investigative Tools

Ironically, the same software that has made it so easy to download copyrighted music makes it exceedingly easy for copyright holders to identify the people who are downloading and sharing their content. When a P2P software user searches for a particular title—Lady Gaga's "Telephone" or James Cameron's *Avatar*—the software identifies all the other active P2P users in the network who have that content stored in their shared directory, and produces a list for the searcher. The list includes the names of the files being shared, the names of the users sharing the files, and the Internet Protocol (IP) address of the sharers' computers. To download the song or movie, the searcher simply clicks on the file name, and the software retrieves it. When the file has been downloaded, the searcher becomes a sharer of that file.

Copyright holders like record labels and movie studios now employ teams of investigators who search P2P networks for their protected content. When they find someone sharing copyrighted material, they record the username, which may not tell them much, and the IP address, which can actually tell them quite a bit.

Every device that connects to the Internet is assigned an IP address by an Internet Service Provider (ISP), the company that provides you and your family with online access. The IP address can be static (when a device, like a router, is connected to the Internet all the time), or it can be dynamic (the device connects to the Internet for a period of time and then logs off). In either case, the ISP can and usually does maintain a record of the IP addresses that have been used by a particular subscriber over the past several months. If a copyright holder sees its content being shared illegally by someone at a particular IP address, it can look up the owner of the address (you can test this yourself with online services like ip-lookup.net or whois.net). The owner will typically be an ISP, and the copyright holder will then issue a subpoena to the company asking for the identity of the subscriber using the IP address on the date and time the infringement was detected. If the IP address is being used by a single computer to connect to the Internet, then the investigation is relatively simple. If, on the other hand, the IP address is being used by a broadcast device such as a home router, then the challenge will be identifying who has connected to the router and downloaded content (which is the chief reason you should make sure your home router is password protected).

Once the copyright holder has identified the name of the subscriber (typically either you or your spouse), a demand letter will be sent seeking payment for the alleged infringement.

The letter typically lists the content the copyright owner believes you or your child downloaded illegally and proposes settlement fee. If you decline to pay the amount set forth in the letter, the copyright owner can file an infringement lawsuit in federal court.

4. Consequences

There are two types of proceedings to enforce a copyright. First, the copyright owner can file a civil suit seeking monetary damages. Under the terms of 17 U.S.C. § 504, the copyright owner can recover either the actual damages cause by the infringement, or so-called statutory damages, which can range from $750 to $30,000 per infringement (so, for instance, a record company could claim up to $30,000 for each song that was illegally downloaded—and they usually do). If the copyright owner succeeds in persuading the court that the infringement was "willful"—i.e., the downloader knew what he or she was doing and was aware it was illegal—then the statutory damages can increase to $150,000 per violation.

There are three different situations in which your child could also be charged with criminal copyright infringement: (1) if he willfully infringes a copyright for personal financial gain (e.g., selling bootleg copies of music or software); (2) if he willfully copies or distributes one or more copyrighted works with a total retail value of at least $1,000; or (3) if he willfully takes a copyrighted work made for commercial distribution and makes it available on a publicly accessible computer network (such as a P2P network or torrent site).

Conviction for criminal copyright infringement can result in prison sentences of up to five years for a first offense. So far, however, there have been relatively few criminal prosecutions of individuals for copyright infringement; criminal charges are generally reserved for businesses that have stolen and redistributed copyrighted material on a grand scale. But given the cost of a legal defense and the potential fines, even a civil suit for copyright infringement can really ruin a family's day. By far, the most well-known civil infringement case was the one brought by the Recording Industry Artists of America against a Minnesota woman named Jammie Thomas-Rassett, who was found to have downloaded and redistributed just twenty-four copyrighted songs. The jury ordered her to pay *$1.9 million* in damages, but U.S. District Court Judge Michael Davis reduced the amount of damages to "just" $54,000, or $2,250 per song, and gave the parties a week to accept the reduced award or structure a settlement. After the parties were unable to agree on a settlement, a third trial was held, and the jury set the amount of damages at $62,500 per song, or a total of $1.5 million. Thomas-Rassett and her attorneys have appealed the decision, arguing that the amount of the award is unconstitutional. Regardless of the outcome of the appeal, Thomas-Rassett has undoubtedly incurred tens, if not hundreds, of thousands of dollars in legal fees.

Recent reports make it clear that the amount of copyright litigation is rapidly increasing. Between January 2010 and January 2011, more than 100,000 peer-to-peer users have been sued for downloading and distributing protected content. While that's still only a tiny fraction of the millions of file sharers, it just takes one copyright infringement lawsuit to drain your pocketbook.

5. Prevention

The first and most effective way to prevent copyright infringement is to teach your children from an early age that it is both morally wrong and illegal to download software or music without paying for it. Part of the lesson, once again, is modeling good behavior yourself. It's difficult to make a compelling case for electronic honesty if your own iPod or computer is filled with pirated music.

Second, it's important to learn enough about the various types of file-sharing programs to be able to determine whether your child is using one. (A list of the leading P2P and torrent programs along with information about where to look for them on your child's computer can be found at www.cybertraps-fortheyoung.com). If your child has installed P2P software, ask why, and how it's being used. There are legitimate uses for P2P software—the legal distribution of public-domain content, for instance, and some artists use these networks to release promotional materials—but the vast majority of material circulating on P2P networks belongs to someone else, and should not be downloaded. It's also important to note that P2P networks are the leading distribution mechanism for child pornography.

Finally, make sure you pay attention to your child's music collection. If your son or daughter has an iPod stocked with far more music than you've given as gifts or that he or she could afford to purchase, it's reasonable to assume at least some of that music has been illegally downloaded. Every unauthorized download increases the likelihood that your child (and you) will attract the attention of record companies intent on enforcing their copyrights.

High-Tech Cheating in Schools

Thanks to the small size and powerful capabilities of today's electronic devices, there is a growing epidemic of academic dishonesty at virtually all levels of the American educational system. Of course, cheating in school was a problem long before the introduction of portable technology. But there is no question that the ubiquity of camera phones, mp3 players, and other devices has increased the level of temptation for students who are facing an increasingly competitive school environment. The only silver lining is that the sheer creativity of some of the methods of cheating bodes well for American entrepreneurship, if not professional rectitude. (For a disturbing peek into the epidemic of cheating, type the phrase "how to cheat" into YouTube and look at some of the examples that pop up.)

1. The Offense

Academic dishonesty is generally defined as cheating in connection with some academic exercise or requirement. The classic example of academic dishonesty is sneaking a "cheat sheet" or "crib notes" into an exam, but using paper to cheat is definitely old school.

2. The Scope of the Problem

The most recent comprehensive study of high-tech cheating was conducted in 2009 by Common Sense Media, and the results were disturbing. Among the most important findings:

- 35 percent of teens with mobile phones admitted to using their phones to cheat. 65 percent of all teens said someone in their school had used a mobile phone to cheat.
- Only 4 percent of teens said storing information in a mobile phone for use on a test is a serious cheating offense. 23 percent of teens said it wasn't cheating at all.
- A similar number—20 percent—said it's not cheating to text test answers to a friend. A much higher percentage of students, however—45 percent—were at least willing to acknowledge the seriousness of such an offense.

One thing made abundantly clear by the Common Sense Media survey is that parents are in a profound state of denial when it comes to electronic cheating. Three out of four parents said they believe mobile phones are used to cheat in their children's schools, but only a scant 3 percent believed their own children had done so.

So, exactly how are children using their mobile phones to cheat? Here are the top five methods uncovered by the survey:

- Using a mobile-phone camera to take a picture of a test and send it to friends
- Searching the Internet for answers to test questions
- Texting friends to warn them about pop quizzes
- Texting friends for help during a test
- Storing information on a phone for use during a test (the old crib sheet, updated for the twenty-first century)

The survey concentrated on mobile-phone cheating, but phones aren't the only devices kids can use for nefarious purposes in the classroom. An mp3 player like the iPod can store more than just music; kids can record themselves reading class notes to play back during a test, or even videotape themselves, with close-ups of notes and textbook information. Programmable calculators can store formulas, notes, and test answers. Even the simplest mobile phone, one without Internet access or photo storage, can still store answers in its contact list. If your child's contact list contains the names of the U.S. Presidents and their phone numbers look like the dates they served in office, you should probably start asking questions.

3. Detecting High-Tech Cheating

The investigation and detection of electronic cheating takes place in two locations: at the school itself and, to a lesser degree, at home. For obvious reasons, teachers and school administrators are the first responders when it comes to academic dishonesty. They are the ones who monitor tests and evaluate the results for inconsistencies, and they have years of institutional experience in ferreting out cheaters. If your child is cheating, it will almost certainly be brought to your attention by a teacher or school administrator.

It is worth noting, incidentally, that school investigations into cheating are among the leading causes (along with
sexting) for the search and seizure of student electronics. A
complete discussion of the constitutional issues surrounding
school searches could fill its own book, but suffice to say that
your child is protected by the Fourth Amendment from unrea-
sonable searches and seizures by school officials. Nevertheless,
federal courts have tended to give those officials a fair amount
of latitude when it comes to enforcing school rules.

Electronic cheating is more difficult to detect at home because so much of it resembles legitimate studying. It's fairly
typical, for instance, for students to retype their notes while
preparing for an exam. The problem arises if those notes are
then stored on a phone for use during the test—and since it
can take just seconds to transfer a file from the computer to a
phone or mp3 player, the chances are slim you'd actually see
that transfer taking place. Given the capabilities of the latest
generation of smartphones, your child could even store test answers on a cloud site like Dropbox, then access them during
the test. In fact, files for each chapter of this book were written using text-editing software that saved each file directly to
Dropbox—meaning I could open the documents on my iPhone
and make edits and corrections on the go. The point is there
are myriad ways for content to be stored remotely and accessed
using a mobile device.

If you routinely examine the contents of your child's
phone, you might see images or files that give you grounds to
ask further questions: photos of exams, for instance, that your
child texted to friends; notes in another child's handwriting;
or copies of answers in a text book. Your child may tell you she
stores notes on her smartphone so she can study in a variety
of locations, and that she has no intention of looking at those
notes during a test. And that may actually be the truth. But
keep in mind that rather large gap between the percentage of

students who admitted to cheating and the percentage of parents who believed their children were capable of such a thing. It's natural to want to trust your children, but trust needs to be earned. It doesn't hurt to pair it with healthy skepticism.

4. Consequences

As with plagiarism, the consequences of high-tech cheating will vary with the severity—or frequency—of the offense. A student caught cheating on an exam would almost certainly receive zero credit. If it's a key midterm or final, that zero could be enough to significantly pull down a student's final grade, even to the point of landing that student in summer school. Since cheating is also a violation of every academic honor code, there may be additional sanctions as well: removal from teams or extra-curricular activities, suspension, and, in sufficiently severe cases, expulsion. Regardless of the precise level of punishment, cheating is an indelible black mark that will blot your child's academic career.

5. Prevention

Like so many of the cybertraps discussed in this book, the most effective prevention is early prevention. From a young age, children need to be taught the importance of honesty, both in general and with respect to schoolwork. This is yet another area where children can benefit from having good behavior modeled for them. If they see their parents and older siblings acting with integrity, they're more likely to do so themselves.

As the Common Sense Media survey results demonstrate, the importance of academic honesty is a lesson that needs to be reinforced as children get older and the pressures for success grow more intense. It can be difficult for children to make the connection between their actions on a test—particularly if a lot of their classmates seem to be cheating—and the consequences in the real world. The temptation to cheat can be particularly strong when parents make good grades the sole measure of a child's success.

Parents should not assume that children will simply figure out the consequences of cheating on their own. All too often, the message children get from their peers, the entertainment industry, and even the evening news is that cheating is either a minor transgression or a legitimate tool for competing with harder working or brighter classmates. Moreover, since a significant amount of school cheating does go undetected, many students think they're cleverly beating the system. The challenge for parents is to find a firm but non-confrontational way to explain that cheating may offer instant gratification for a particular test—an academic "sugar high," as it were—but integrity and moral fiber are a better long-term diet.

One way this can be accomplished, both in school and at home, is through the discussion and adoption of an academic honesty code. Many schools around the country require middle- and high-school students to read and sign such a code at the beginning of the academic year, and it can be useful to review it with your child. An even more effective step is to draft an academic honesty policy with your child for use in your home (some suggested themes and statements can be found online at www.cybertrapsfortheyoung.com). Drafting the document together will help your child understand the importance of integrity at school, and will give him or her some ownership of the concept—the final document will be something you've come up with together, rather than something you're enforcing from on high. Most important, creating this document will underscore the message that you consider *how* a grade was earned to be just as important—if not more so—than the grade itself.

Chapter Six
Electronic Harassment, Cyberbullying, and Cyberstalking

Few people look back with fondness on their years in middle school or junior high. Those years tend to be difficult, thanks to riptides of hormones, rapidly shifting relationships with classmates, and increased academic responsibilities. All too often, children respond to the rising pressures by lashing out at others, and in some cases even conduct coordinated campaigns of harassment and bullying against less popular or less accepted peers. If you haven't recently read William Golding's *Lord of the Flies,* pick up a copy at your local library—it's a chilling depiction of what many kids experience in their early adolescence.

At first blush it might seem hyperbolic to compare the typical middle school with an island of stranded children fighting each other for survival. But recent events in western Massachusetts show just how similar those worlds can be.

In the spring of 2009, Carl Joseph Walker-Hoover, a sixth grader at the New Leadership Charter School in Springfield, Massachusetts, hanged himself after being regularly bullied and harassed. In published reports after his death, his mother said that her son had been repeatedly threatened, constantly taunted, accused of being gay, and mocked for his clothing.

Eight months later, at nearby South Hadley High School, a freshman named Phoebe Prince also hanged herself, in a stairwell in her home, after enduring months of harassment from her fellow students. Prince had recently come to the school from Ireland and, according to reports, was verbally taunted, physically assaulted, and electronically harassed on a regular basis. The story is deeply complicated, but much of the harassment was apparently driven by a perception that Prince was interfering with existing relationships in the school. One typical Facebook post from a jealous girl read: *"Know what I hate? Irish sluts."*

Shortly after Prince's death, the local district attorney, Elizabeth Scheibel, filed criminal charges against six South Hadley High students. The alleged crimes included violations of Prince's civil rights through bodily injury (her death by suicide), statutory rape, harassment, stalking, and disrupting a school assembly. The trials of the six defendants, all of whom have pled "not guilty," is slated to begin in 2011.

These two high-profile cases prompted the Massachusetts legislature, in the spring of 2010, to adopt one of the nation's strictest anti-bullying laws. Among other provisions, the law requires every teacher and school staff member to report any incidents of bullying they observe, and mandates that each report of bullying be investigated by the principal or a designated staff member. Ongoing training is required for teachers and staff to help them both identify and prevent bullying, and similar instruction is required of students at every level of the curriculum.

According to the website BullyPolice.org, forty-five states have passed anti-bullying laws, which the site helpfully grades according to comprehensiveness and commitment to anti-bullying education for both children and adults. Passing laws against cyberbullying and harassment, however, is only the first step to solving this growing problem. A solid, long-term commitment to education and training is needed to combat the trend of children using new technology to engage in ages-old bad behavior.

Electronic Harassment and Cyberbullying

1. The Offenses

There is significant overlap between electronic harassment and cyberbullying, but they are distinct criminal offenses and are often treated separately under state law. The National Conference of State Legislatures defines cyberharassment as "threatening or harassing e-mail messages, instant messages, or blog entries, or websites dedicated solely to tormenting an individual." Cyberbullying, meanwhile, is generally considered to be a subset of the broader category of cyberharassment, occurring "between minors within the school context."

The recent Massachusetts law offers a good illustration of how legislatures are struggling to encompass all the changes in technology children can use to bully others. The law contains a definition both for "bullying" and for "cyberbullying":

> "**Bullying**," the repeated use by one or more students of a written, verbal or electronic expression or a physical act or gesture or any combination thereof, directed at a victim that: (i) causes physical or emotional harm to the victim or damage to the victim's property; (ii) places the victim in reasonable fear of harm to himself or of damage to his property; (iii) creates a hostile environment at school for the victim; (iv) infringes on the rights of the victim at school; or (v) materially and substantially disrupts the education process or the orderly operation of a school. For the purposes of this section, bullying shall include cyber-bullying [sic].

> "**Cyber-bullying**," [sic] bullying through the use of technology or any electronic communication, which shall include, but shall not be limited to, any transfer of signs, signals, writing, images, sounds, data or intelligence of any nature transmitted in whole or in part by a wire, radio, electromagnetic, photo electronic or photo optical system, including, but not limited to, electronic mail, internet communications, instant messages or facsimile communications. Cyber-bullying

shall also include (i) the creation of a web page or blog in which the creator assumes the identity of another person or (ii) the knowing impersonation of another person as the author of posted content or messages, if the creation or impersonation creates any of the conditions enumerated in clauses (i) to (v), inclusive, of the definition of bullying. Cyber-bullying shall also include the distribution by electronic means of a communication to more than one person or the posting of material on an electronic medium that may be accessed by one or more persons, if the distribution or posting creates any of the conditions enumerated in clauses (i) to (v), inclusive, of the definition of bullying.

In simpler terms, what the Massachusetts legislature is attempting to do is to create a physically and emotionally safe school environment for all children, and to prohibit a variety of types of abuse that interfere with that goal. The Legislature defines cyberbullying as a subset of the broader category of bullying, consisting largely of abusive acts (bullying) that are committed by electronic means. It will be interesting to see if the Massachusetts legislature (or others around the country) find it necessary to periodically update their laws as new technologies emerge.

2. The Scope of the Problem
A number of surveys have attempted to quantify just how much cyberbullying and cyberharassment takes place in the nation's schools, but those numbers are tough to pin down. Kids are understandably reluctant to admit to doing something that, at the very least, might get them in trouble at school and, at worst, might land them in serious legal hot water.

One of the most commonly cited surveys—conducted in 2004 by i-SAFE America—found that nearly 60 percent of kids in grades four through eight had been the target of a mean or hurtful statement online, while just over half of all students admitted to making such mean or hurtful statements. Of particular concern: nearly 60 percent of kids who were targeted online said nothing about the bullying or harassment to their parents.

Coincidentally, the year of the iSAFE survey, 2004, was also the year in which mobile-phone companies began aggressively marketing to teens. At the time, only one in five teens carried a mobile phone; now that figure is closer to 95 percent. It's not hard to imagine that all those mobile phones have exacerbated what was already a serious problem. It takes mere seconds, after all, for a kid to type out and send a three-word text message that can ruin a classmate's day. *You're so fat! Nobody likes you! Who dresses you? I hate you!* If those seem cruel or hurtful, rest assured that kids are capable of coming up with far worse.

Of course schoolyard bullying has been a problem as long as schools have been in existence. Back when Alexander Graham Bell was experimenting with prototypes of the telephone, the children in his neighborhood were no doubt taunting each other about the quality of their knickers or their poor stickball skills. But our modern-day onslaught of electronic gadgetry has changed the playing field for bullies and their victims. The new technology is fast, and it's effective at reaching a broad audience quickly—a bully's voice is no longer restricted by how loudly he can yell across a playground. Perhaps worst of all, the new technology encourages remote bullying—it's just human nature, after all, that people are more apt to say mean things about someone when they don't have to face the victim of their taunts. The remote quality of electronic bullying also makes it harder for victims to hide from their tormentors. When I was a kid, if someone picked on me, no matter how merciless they might be, I knew at the end of the school day I could find sanctuary at home. Now, thanks to computers and mobile phones, bullies can follow kids home electronically, making their bullying all the more relentless.

One thing that makes cyberbullying such an insidious problem is that there are so many ways to bully someone electronically. Today's technologically savvy bullies are limited only by their own creativity. Just a few examples:

- Insulting Internet polls can be set up on websites like SurveyMonkey.com or MisterPoll.com and forwarded to classmates ("Who's the fattest in our class?" "Hot or Not," "Vote for the Biggest Slut").
- Kids can be subject to abuse in online games, including exposure to hurtful messages or teams of other kids unfairly ganging up on them.
- A bully might try to infect his victim's computer with malicious code attacks, such as viruses or Trojan horses.
- Kids can be signed up, without their knowledge, for inappropriate or even offensive websites.

Bullies also might engage in various forms of identity theft, either making inappropriate statements while posing as the victim in a chat room, or using a victim's name to create an offensive social networking profile. Particularly malicious is when a cyberbully wreaks havoc by proxy, using the victim's name or ID to initiate bad behavior online, which often results in the victim being punished by a duped authority figure.

3. Investigative Tools

Most of us—sometimes from personal experience—know all too well the emotional and even physical consequences that can result from bullying, regardless of whether it occurs online or off. Generally, our first concern is whether our children are being victimized by bullies. However, as parents we also must force ourselves to consider the unthinkable: what if our child is, in fact, the one doing the bullying? As illustrated by the charges levied against the six South Hadley students, if your child *is* bullying the legal and financial consequences can be profound.

The first step in determining whether your child is a cyberbully is to simply pay attention to his or her interactions with friends (and with you). What types of things does your child say about other kids? When your child's friends are around, do they routinely make fun of certain people? Do they tell jokes at the expense of others, or mock particular characteristics of their classmates? If your child has a habit of doing so, or suddenly begins doing so, it could be a first warning sign that abusive behavior is also taking place online.

If you have concerns that your child is cyberbullying other kids, or if another parent or child reports the behavior, you'll need to spend time reviewing your child's electronic communications. This includes the following:

- Reading text messages, either directly on your child's phone or by requesting copies of recent texts from the phone company
- Logging in to the social networking sites your child frequents, like Facebook and MySpace, and reviewing not only his or her public activity, but any person-to-person messages or IM sessions that occurred on the site
- Logging in to your child's e-mail account and reviewing messages and chat logs
- Searching the Internet for your child's name or the name of the alleged cyberbullying victim to see if anything crops up

If another parent or teacher reports that your child is cyberbullying, they'll likely be able to tell you exactly how the cyberbullying occurred (text, e-mail, social networking site, etc.), which will make it easier for you to focus your own investigation. Keep in mind, though, that if your child has been caught cyberbullying using one method, other methods are likely being exploited as well, so don't limit your investigation to one or two online tools.

4. Consequences

Children who engage in cyberbullying or cyberharassment can either be charged under a specific statute (as in Massachusetts) or with a variety of existing state prohibitions against stalking, criminal harassment, identity fraud, hazing, and, more broadly, violating someone's constitutional rights. What specific charges a child might face will depend on the facts of the case and the laws of the jurisdiction in which the crime occurs. In the typical cyberbullying case, of course, both the perpetrator and the victim are in the same jurisdiction, since they're usually in the same school. That means that any potential criminal charges will be governed by the laws of the state in which you and your child live. But don't lose sight of the fact that, thanks to the remarkable communication capabilities of the Internet, kids can now bully each over hundreds or thousands of miles. As I previously noted, forty-five of the fifty states have adopted specific laws targeting cyberbullies. Links to those laws, as well as to other cyberbullying resource sites, can be found online at www.cybertrapsfortheyoung.com.

Currently, there is no federal law dealing specifically with cyberharrassment or cyberbullying among children, though a bill was introduced in January 2009—the Megan Meier Cyberbullying Prevention Act—that would have established a maximum two-year prison sentence for "whoever transmits in interstate or foreign commerce any communication, with the intent to coerce, intimidate, harass, or cause substantial emotional distress to a person, using electronic means to support severe, repeated, and hostile behavior. . ." The bill ultimately stalled in committee, due in part to concerns about its constitutionality under the First Amendment. But given the media attention cyberbullying has received in recent months, it's likely Congress will eventually revisit the issue in one way or another.

It's also important to note that some kinds of cyberbullying—such as impersonating someone else in a chat room or on a social networking site—may result in a child being charged with the much more serious crime of identity theft, discussed in more detail in chapter 9.

5. Prevention

To be fair, teens and preteens say a lot of stupid or thoughtless things—whether online or off—and their relationships with peers can be coarse, crude, and mocking, without actually rising to the level of bullying or harassment. That can make it quite a challenge to figure out which of your child's statements may actually cross the line into legally or morally dubious territory. Since parents obviously can't monitor every statement that comes out of their children's mouths (or electronically, through their fingers) the single most important tool for preventing cyberbullying is education—teaching children about the harm that can result from thoughtlessly cruel or malicious speech.

The most important part of that education takes place at home, and should begin well before a child has any exposure to the Internet or electronic communication—though, as we saw at the beginning of this book, access to the digital world is occurring at an earlier and earlier age. Teach your child to be kind and respectful to others, and to imagine how it would feel to be on the receiving end of hurtful texts or online posts. The "golden rule" is golden for a reason.

If you're looking for a clear and simple message for young children, a good starting place is the anti-cyberbullying campaign of the National Crime Prevention Council: Don't Write It. Don't Forward It. Stop Cyberbullying. The campaign, located online at http://www.ncpc.org/cyberbullying, has a variety of useful resources for parents.

Particularly when children are young and still figuring out the appropriate boundaries of online behavior, it may be a good idea to restrict their use of electronics to common areas of the home—the living room, family room, or kitchen—and to exercise a high level of surveillance. Few things are a better deterrent to electronic misbehavior than a parental audience.

As children get older and begin spending more time interacting with classmates and friends online, the warnings about the consequences of cyberbullying should become more detailed and frequent. If your child's middle school does not discuss appropriate online behavior as part of its regular curriculum, work with the administration to develop an appropriate program. Invite the school's student resource officer or another police representative from the school to talk with students about the consequences of harassing people or assuming their identities online. Lastly, don't overlook the importance of networking with other parents to set up an early warning system for potential trouble. The child you keep out of the legal system may be your own.

The Perils of Obsession: Cyberstalking

There is a point at which an individual's desire to hurt or harass another person online crosses the line from cyberbullying to cyberstalking. We don't typically think of kids as stalkers, let alone cyberstalkers, but teen cyberstalking does happen, and the consequences can be severe.

Three years ago, for instance, a student at Providence High School in Charlotte, North Carolina, was charged with cyberstalking after setting up a website that suggested a male teacher was a pedophile. Under North Carolina law, cyberstalking is a Class 2 misdemeanor, punishable by up to thirty days in jail. Four other students also were punished under the school's code of conduct, which prohibited the distribution of "any inappropriate information, relating in any way to school issues or school personnel . . . from home or school computers."

More recently, in the spring of 2010, a seventeen-year-old Florida student named Shannon Mitchell was arrested and charged with aggravated cyberstalking for posting messages about another girl on an adult website, where she also posted the girl's phone number. The victim's family told reporters she received dozens of sexually explicit phone calls from visitors to the site, asking for her address and trying to set up physical encounters. Under Florida law, aggravated cyberstalking is a third-degree felony, but since Mitchell is a juvenile, it's likely her case will be diverted to juvenile court. If she were over the age of eighteen, however, she would face up to five years in prison and a fine of up to $5,000.

1. The Offense

Al Gore's invention of the Internet has become something of a national punch line, but there's no question that Gore deserves credit for recognizing the growing role of the Internet in stalking cases. On February 26, 1999, then-Vice President Gore asked Attorney General Janet Reno to study cyberstalking and prepare a report on how to address the problem.

That Justice Department report—titled "Cyberstalking: A New Challenge for Law Enforcement and Industry"—sought to define the new crime:

> Although there is no universally accepted definition of cyberstalking, the term is used ... to refer to the use of the Internet, e-mail, or other electronic communications devices to stalk another person. Stalking generally involves harassing or threatening behavior that an individual engages in repeatedly, such as following a person, appearing at a person's home or place of business, making harassing phone calls, leaving written messages or objects, or vandalizing a person's property.

Since the report, thirty-five states and U.S. territories have adopted laws that prohibit cyberstalking. In some states, like New Jersey and New Mexico, the prohibition is inferred from existing laws against stalking, which prohibit the use of any "device" to commit the crime. Most experts interpret that language to include computers, mobile phones, and other electronic gadgets. Other states, meanwhile, have adopted specific cybstalking statutes. In North Carolina, for instance, several specific kinds of cyberstalking are now explicitly illegal:

1. Use in electronic mail or electronic communication any words or language threatening to inflict bodily harm to any person or to that person's child, sibling, spouse, or dependent, or physical injury to the property of any person, or for the purpose of extorting money or other things of value from any person.

2. Electronically mail or electronically communicate to another repeatedly, whether or not conversation ensues, for the purpose of abusing, annoying, threatening, terrifying, harassing, or embarrassing any person.

3. Electronically mail or electronically communicate to another and to knowingly make any false statement concerning death, injury, illness, disfigurement, indecent conduct, or criminal conduct of the person electronically mailed or of any member of the person's family or household with the intent to abuse, annoy, threaten, terrify, harass, or embarrass.

4. Knowingly permit an electronic communication device under the person's control to be used for any purpose prohibited by this section (N.C. Gen. Stat. §§ 14-196.3).

There's no hard and fast rule for when cyberbullying crosses the line into cyberstalking, and many states haven't bothered to create a separate statute for the crime. In those states that have created a statute to punish cyberstalking, the emphasis appears to be on the severity of the emotional harm to the victim (such as North Carolina's use of the word "terrify") and the frequency and persistence of the abusive messages. Another factor that sometimes comes into play is whether the perpetrator has made threatening or abusive messages in a variety of different media (cell phone, text messages, email, social networks) or on numerous sites.

2. The Scope of the Problem
Unlike the crime of cyberbullying, where rates of child victims and child perpetrators are roughly equal, far more children are targets of cyberstalking by adults than perpetrators themselves. However, there have been a handful of instances in which students have been charged with cyberstalking.

For instance, in 2007, Tyler Yannone and Lauren Strazzabosco, two students at Mooresville High School in Mooresville, North Carolina, were charged with cyberstalking after they created a MySpace profile depicting a school administrator as a pedophile. The two were arrested under North Carolina's cyberstalking law and charged with misdemeanors.

In the spring of 2009, Ahn Nguyen, a student at Knightsdale High School in Knightsdale, North Carolina, was charged with cyberstalking after posting a video to his Facebook page in which he made violent threats toward a student who had earlier compared him to the Virginia Tech shooter. After Ngyuyen's arrest, police searched his home and seized a variety of electronic equipment, including cameras, computers, memory devices and a mobile phone, all of which police suspected might contain evidence of his online aggression.

Then, just a few months ago, Tony Bloomberg and Matthew Nanney, students at Pamlico County High School in Bayboro, North Carolina, were found guilty of cyberstalking after they created a fake Facebook page in their principal's name, using the page to criticize other students in the school. The two were each sentenced to forty-eight hours of community service and made to write letters of apology to named students. The judge did grant the two some leniency, agreeing to expunge the crime from their records if they could complete their community service within six months of sentencing.

It may appear that North Carolina is the epicenter of teen cyberstalking, but a better explanation is that officials have been particularly aggressive in pursuing cyberstalking charges against teens. There have been other student cyberstalking cases reported around the country, and in other nations as well. For instance, in early 2011, a fifteen-year-old Rhode Island girl was charged with cyberstalking after she created a false Facebook page with a picture of a bloody severed foot, which she used to mock a freshman girl who was born with a partial left foot. Police used six search warrants to track the sophomore girl to her home, and seized three computers from her house as part of the investigation. Coincidentally, on the same day, two Lafayette, Louisiana high school students were charged with cyberstalking after using Facebook to repeatedly bully and harass a younger female student.

Given the growing ubiquity of social networks and mobile devices, cyberstalking is a crime that is likely to grow more common unless parents and school officials intensify their educational and enforcement efforts.

3. Investigative Tools

In terms of your child's online behavior, there's not a lot of distinction among cyberharassment, cyberbullying, and cyberstalking. One important distinction, however, is the amount of time required to commit each offense. Cyberbullying or cyberharassment can occur with little more than a one-off text or status update, while cyberstalking typically requires a lengthy and concerted campaign of online activity. As a result, your first investigative tool is the clock: just how much time are your children spending online, and how are they using that time?

Beyond that initial level of inquiry, ferreting out cyberstalking requires the same investigative tools discussed earlier in this chapter: awareness of your children's interactions with their classmates, conversations with your children about their on-line activities, a reasonable level of supervision of their online communication, and networking with teachers and other parents.

The key emotional difference between cyberbullying and cyberstalking is the quality of obsession. While cyberbullying is no doubt mean and hurtful, it can appear relatively casual, and is often short lived. With cyberstalking, however, there is an element of fixation or determination that, if detected, might signal deeper emotional issues in a child. It can also lead to more serious legal consequences. While it is important not to overreact to childish efforts at humor or even jokes that are merely in bad taste, it is worth remembering just how quickly a teen or preteen can use the Internet to transform a childish hatred of a particular teacher into a Facebook group titled "Mr. Jones Is a Racist and Fondles His Students."

4. Consequences

Due to the more severe harm usually contemplated by cyberstalking statutes, a child charged under such laws will typically face harsher penalties, including increased fines and longer potential prison sentences than those typically assessed in cyberbullying cases. The more severe nature of this crime also makes it less likely that a prosecutor will choose to divert a case to juvenile court, or that a judge will impose a sentence of probation, community service, or similar non-prison option. But the prosecution and outcome of every criminal case depends on its unique blend of the facts, the relevant laws, and the people involved. If your child is charged with cyberstalking, it is critically important that you hire an attorney licensed in your jurisdiction who is not only experienced in criminal defense, but either has or can hire the technical expertise to properly analyze the facts and issues in the case.

It's worth noting, incidentally, that parents also can find themselves in trouble under the North Carolina statute (or similar laws) if they are found to be in control of the device used to cyberstalk someone. Whether that refers to actually physical control or financial control remains an open legal question.

5. Prevention

As with cyberbullying, the prevention of cyberstalking needs to begin long before kids are online, as part of a larger lesson about how others should be treated. Many families may feel uncomfortable discussing cyberstalking in much detail with elementary-aged children online for the first time, but by the time a child hits middle school, it should certainly be a topic of discussion.

Given the severity of cyberstalking and the potential legal consequences, it can be useful to ask student resource officers or other law enforcement officials to address the issue with middle and high school students. Since one-time lectures are rarely effective for teens and preteens, parents, schools, and law enforcement should collaborate on ways to integrate cyberbullying and cyberstalking into the regular school curriculum.

The bulk of prevention, however, will need to take place at home. Most children are aware of at least *some* of the potential punishments for cyberstalking and will presumably try to hide such behavior from their parents, so it may not be an activity you can uncover through a casual review of your child's text messages or e-mail account. But there are usually warning signs: offhanded comments expressing hatred or intense dislike for a particular person or group of people, heightened cliquishness, obsession with electronics and social networking tools, abnormal secretiveness, and so on. There may also be alerts from the school administration or your friends and neighbors that there's an ongoing problem. But don't hold your breath for a clear "I'm cyberstalking Sue today" message from your child.

If you do think your child might be cyberstalking someone or collaborating with someone else in doing so, then begin investigating your child's electronic activity immediately. Talk with him about his online activity and interactions with his peers. Reach out to your network of friends, parents, and teachers to see if there any reports or even rumors of online harassment or cyber abuse. Conduct regular (but unpredictable) physical inspections of your child's phone and computer (recent calls, text messages, social network updates, photos, e-mails, etc.).

Some of those steps (apart from face-to-face communication with your child) can be accomplished through the installation of surveillance software on your child's computer. If installed properly, such software can reduce or eliminate the need to physically inspect the computer, since the relevant information can be accessed remotely. See appendix A for a more detailed discussion of the various options that are available for personal computers.

Unfortunately, there are fewer good options for electronically monitoring online activity that takes place on mobile devices. That may change, as parents become more aware of the computer-like capabilities of smartphones and create a demand for such software. For the time being, however, effective monitoring of mobile devices requires actual hands-on inspection.

Chapter Seven
Electronic Defamation and Invasions of Privacy

Defamation

The concept of defamation has been part of Western European law for more than two millennia. The Romans awarded compensation for the public disparagement of an individual's reputation, provided there was no truth to the allegations. The concept later found its way into English common law, and eventually to law in the United States.

Defamation is actually an umbrella term for two specific types of behavior: libel and slander. Libel encompasses untrue *written* statements that harm an individual's reputation, while slander refers to *spoken* (or other "transitory") statements. Traditionally, libel was considered the more serious offense—written defamation being seen as more permanent—but over the years, particularly since the development of the Internet, the distinction between them has narrowed significantly. These days, many writers use the two terms interchangeably.

The reason this concept is important to understand is that there are so many different ways children can express themselves online, and so many people ready to take offense at what they see or read (which is not to say children are incapable of broadcasting patently offensive things offline). Damages awarded in defamation suits can be staggeringly large, and even if no judgment is awarded, the cost of defending a defamation suit can itself be crippling to individuals or families.

Libel and Slander

1. The Offense

Defamation can be prosecuted as either a criminal or civil offense, though civil suits for libel and slander are far more common that criminal suits. Regardless of whether the offense is written or spoken, a civil claim for damages rests on the same four elements. First, the statement in question must be a "non-speculative" one with the potential to injure a person's reputation and expose him or her "to public hatred, contempt, ridicule, or degradation" [*Phipps v. Clark Oil & Ref. Corp.*, 408 N.W.2d 569, 573 (Minn. 1987)]. Non-speculative here means the statement can't simply be an opinion (i.e. "Tina is ugly") but rather must be an assertion of fact that can be proved as true or false (i.e., "Tina regularly French kisses her poodle."). Second, the statement must have been spoken (or published) to third parties. Finally, the speaker (or writer) had to know—or at least *should* have known—that the statement in question was untrue.

There are two other important things to understand about defamation. The first is that truth is a complete defense to the claim—in other words, if a defendant can show the statement was true, regardless of how hurtful or mean-spirited it was, there can be no damages awarded for defamation. Of course that might be cold comfort if you're staring at a pile of legal bills incurred while your attorney proved the truth of your son's blog posts. Still, it beats getting stuck with a monstrous fee for damages.

The second thing to keep in mind is that some statements are so vile and offensive that they constitute what is known as "defamation per se." In these cases, the plaintiff does not have to actually prove that he or she incurred damages; the defamation is so awful that damages are simply assumed. Classic examples of defamation per se include allegations of serious sexual misconduct (rape, child sexual abuse), grave criminal conduct (murder, kidnapping), or a "loathsome disease" (traditionally leprosy or venereal disease, but today this category might include AIDS).

By the middle part of the last decade, prosecuting individuals for *criminal* libel seemed to be dying out. As of 2006, two-thirds of the states had repealed their criminal defamation laws or else seen them overturned by courts on First Amendment grounds (the theory being that the government has no business punishing you for what you say about someone else). However, with the growth of the Internet and the rise of social networking sites, the concept of criminal defamation is beginning to once again gain favor. In large part, support stems from the profoundly serious consequences that can result when maliciously false statements are spread quickly and widely online.

The language used in the Kansas statute on criminal defamation offers a good illustration of how these laws are typically structured:

> Criminal defamation is communicating to a person orally, in writing, or by any other means, information, knowing the information to be false and with actual malice, tending to expose another living person to public hatred, contempt or ridicule; tending to deprive such person of the benefits of public confidence and social acceptance; or tending to degrade and vilify the memory of one who is dead and to scandalize or provoke surviving relatives and friends [Kan. Stat. Ann § 21-4004 (2005)].

Criminal defamation is typically treated as a misdemeanor, though three states—Colorado, Florida, and Michigan—consider it be a felony. Conviction can result in imprisonment for six months or more (Alabama provides for hard labor), and fines of varying amounts.

Another source of possible punishment for statements made online comes from colleges and universities, many of which have prohibitions against hate speech. The language contained in the Emory University hate speech code is characteristic:

> Discriminatory harassment includes conduct (oral, written, graphic, or physical) directed against any person, or group of persons, because of their race, color, national origin, religion, sex, sexual orientation, age, disability, or veteran's status, and that has the purpose or reasonably foreseeable effect of creating an offensive, demeaning, intimidating, or hostile environment for that person or group of persons.

Large numbers of colleges and universities have adopted such codes over the past several years. Those adopted by public universities have been challenged on First Amendment grounds and frequently overturned. However, private colleges like Emory aren't bound by First Amendment concerns and are free to adopt whatever codes of conduct they see fit. Many students have been disciplined at colleges for what they've posted online, with punishments ranging from suspension to outright expulsion.

2. The Scope of the Problem

There are no specific statistics regarding the number of civil and criminal defamation cases involving children, but anecdotal evidence makes it clear these cases are a rising phenomenon, thanks primarily to the growth of social networking sites and the ease with which the sites can be used to harass other students, teachers, and administrators. So far, at least, plaintiffs have had a difficult time persuading courts that their claims are valid, but that may well change in the coming years. A couple of examples help illustrate how online statements can lead to both civil litigation and criminal prosecution.

In the fall of 2006, Anna Draker, an assistant principal at Clark High School in San Antonio, Texas, was informed that someone had set up a MySpace page using her name. The primary purpose of the page was to suggest, both indirectly and directly, that Draker was a lesbian, a claim which was untrue. There were also a number of comments left on the page by students of the school—comments which, according to Draker's complaint, were "suggestive, lewd, and obscene."

With the assistance of MySpace, Draker had the page removed and obtained information that led to the identification of the responsible students. Those students, Ben Schreiber and Ryan Todd, evidently created the page as retaliation for disciplinary action that had been taken against them. Draker filed suit against both students for "defamation and libel per se," and against both sets of parents for negligence and negligent supervision. She sought an award for a variety of damages, including lost wages, emotional distress and mental anguish, and court costs. Ben Schreiber was also prosecuted under Texas criminal law for retaliation and fraudulent use of identifying information.

Ultimately, in the civil case, the students and their parents were able to persuade the trial court that the "exaggerated and derogatory statements" on the website were speculative—that is, they couldn't be proven or disproven—so one of the key elements for a defamation claim was missing. A Texas appeals court affirmed that decision on August 13, 2008. Since Schreiber was a juvenile, the outcome of the separate criminal proceeding is unknown.

In the spring of 2009, Denise Finkel, a graduate of Oceanside High School in Oceanside, New York, sued four high school classmates for defamation after they set up a Facebook group that allegedly held Finkel up to "public hatred, ridicule, and disgrace." Though the site didn't name her specifically, Finkel said it was intended to imply that she engaged in bestiality, was an IV drug user, and suffered from AIDS. Finkel also sued her classmates' parents for negligent supervision of a "dangerous instrument" (i.e., a computer), as well as Facebook, for hosting the group's page. Altogether, Finkel sought $3 million in damages.

In October 2009, a New York trial judge dismissed the claim against Facebook, saying the company is protected from liability for content posted by its users, under the Communication Decency Act. A year later, the trial court dismissed the case against Finkel's classmates, saying that although the Facebook posts "constitute[d] evidence of adolescent insecurities and indulgences, and a vulgar attempt at humor," they did not contain the type of non-speculative statements needed to meet the defamation standard. The court also dismissed the claims against the parents, holding that under New York law computers are not "dangerous instruments."

Obviously, these particular lawsuits were not successful, but they offer some insights into the types of civil claims that can be brought. Criminal charges are rather less likely, but given the broad language used in many state criminal defamation statutes, it's not out of the realm of possibility that a prosecutor might try to make an example of a teen who makes one or more particularly outrageous statements online that "tends to expose any individual or any religious group to hatred, contempt, ridicule, or obloquy" (illustrative language from the Florida criminal defamation statute, 836.11).

3. Investigative Tools

Having full friend status with your child on any social network is an important step in detecting possible problems, as is subscribing to his or her Twitter feed and RSS feeds for blogs, checking his or her text messages, and occasionally reviewing e-mails. If you're paying attention, you may well get a hint that something unusual is taking place; it's the rare child who can entirely filter or segregate his or her conversations. But detection may not be simple: most kids are aware that this kind of behavior is reprehensible, so they're likely to go to great lengths to make sure their parents won't uncover it. That may include setting up secret e-mail accounts or new social-networking IDs to keep their activities hidden.

The key lesson: if you're concerned that your child is actively hiding Internet activity from you, because you see significant behavioral changes, open antagonism to classmates or teachers, or an unusually high level of evasiveness, then it's time to consider a more active and intense level of supervision.

Keep in mind, however, that there is some legal risk in capturing information using surveillance software (for more detailed information, please see appendix A). If you and your child are sued for defamation—particularly if the suit includes a related charge for negligent supervision—you could be required to produce the record you've created of your child's online activity during the discovery phase of the suit. That information could, unfortunately, prove quite useful to a plaintiff in presenting his or her case to a jury. Moreover, a plaintiff might be able to argue that the activity log shows that you were aware of the defamatory activity and that you did not take adequate steps to promptly stop it.

Those potential harms are relatively remote, however, and don't really come close to outweighing the benefits of having more detailed information about your child's online activity. There's no hard and fast rule about when you should absolutely install surveillance software, but it's certainly better to be aware of potential problems before they become actually crimes.

4. Consequences

While it may look as if kids get a pass when it comes to defamation claims, it's important not to draw the wrong conclusions from the limited pool of teen defamation cases discussed above. It may be more difficult to bring a successful defamation claim against students, since judges know—many from personal experience—that kids are prone to saying stupid things, exaggerating for effect and not stopping to think about the implications of what they say. But there have been several successful defamation suits brought against adults for statements made online, often resulting in enormous damage awards. For instance, a British Columbia court assessed $179,644.50 against a man who published comments about another man online that he was a "stalker, abuser, harasser, criminal, evil, liar, killer, sexual predator, pervert, pedophile, coward, manipulator and hate monger who threatens others with death and violence, betrays confidences and is untrustworthy, is dangerous, abuses, impregnates and abandons women and publishes their personal information." And a decade ago, a man won a $40,000 settlement from an Internet service provider (along with $600,000 in legal fees) after claiming he'd been defamed in a chat room hosted by the ISP.

It's nearly inevitable that someone will eventually bring a successful claim against a teenager. Also, keep in mind that even those claims that haven't succeeded—like the ones brought by Draker and Finkel—have cost defendants tens of thousands of dollars in legal fees, and have left the adolescent defendants with widely publicized black marks that, fairly or not, are now a permanent part of their electronic resumes.

On the criminal front, defamation is generally considered to be a misdemeanor or, at most, a low-level felony. As a result, the most common penalties are probation, community service, or a written letter of apology. Fines typically range from as little as $100 to a maximum of a few thousand dollars, while incarceration is generally six months or less. Fortunately, the First Amendment implications of criminal defamation laws make these an exceedingly rare type of prosecution.

5. Prevention

The best tools for preventing possible defamation by your child are education, communication, supervision, and networking—the central themes that run throughout this book.

The education piece is straightforward: children need to be taught from an early age that using any Internet communication tool to say mean or malicious things about another person is a violation of your home code of conduct, his or her school's code of conduct, and—perhaps most importantly—the law. There are few mediums for which the saying "If you don't have anything nice to say, don't say anything" is more applicable than the Internet. Children have a tendency not to think of the consequences of comments they post to Facebook, or send in an e-mail or text, and your first job as a parent is to make them understand that those consequences can be very real.

The next effective tool is an ongoing conversation with your child about his or her online activity. Listen to how your child characterizes his or her interactions with peers and teachers. Is your son angry at a particular teacher or administrator? Does your daughter speak with special venom about one or more classmates? It's natural, of course, for children to dislike certain teachers, or to become irked with their peers—teenagers can often be best friends one week and enemies the next—but you should emphasize to your child that broadcasting those feelings to a larger audience can be hurtful and inappropriate—even illegal.

In addition to the emotional and social issues, pay close attention to your child's growing online skill sets. Is your son a Facebook power user? Does your daughter have the know-how to create a defamatory website? Do your children have friends with those skills? As the parent of teens, I know that these are difficult measures to carry out, but with time and patience we can develop a sense of just how adept our children are with Internet tools.

As our discussion progresses—and, yes, there are even more serious cybertraps your child can fall into—you may conclude that the risks are so severe as to warrant some level of electronic monitoring. Not surprisingly, many parents choose to conduct more aggressive surveillance when their children are younger and more impetuous, and ease off as their children mature and demonstrate more responsibility in their online behavior. Whether surveillance is necessary will depend on the assessment of your child's current emotional state, social interactions, and his or her computer skills.

A final preventative measure, networking, is really a carryover from the issues of cyberbullying and cyberstalking already discussed. In all cases, networking with other parent and school officials can create a crucial early-warning system, as well as a means for conflict resolution when problems do arise. If the school doesn't have a specific code of online conduct, a collaborative group of parents and teachers can help create one, and can organize informational meetings for children and parents.

Privacy Torts and Crimes

One hundred twenty years ago, two Boston law partners, Louis Brandeis (later a U.S. Supreme Court justice) and Samuel Warren, wrote an article for The *Harvard Law Review* titled "The Right to Privacy." In it, the two lawyers argued that thanks to new technologies like the camera and high-speed printing presses, it was time for the courts to recognize a new claim for damages resulting from the violation of a person's right to privacy. Their argument in support of this new right was remarkably prescient:

> Of the desirability—indeed of the necessity—of some such protection, there can, it is believed, be no doubt. The press is overstepping in every direction the obvious bounds of propriety and of decency.

Gossip is no longer the resource of the idle and of the vicious, but has become a trade, which is pursued with industry as well as effrontery. To satisfy a prurient taste the details of sexual relations are spread broadcast in the columns of the daily papers. To occupy the indolent, column upon column is filled with idle gossip, which can only be procured by intrusion upon the domestic circle. The intensity and complexity of life, attendant upon advancing civilization, have rendered necessary some retreat from the world, and man, under the refining influence of culture, has become more sensitive to publicity, so that solitude and privacy have become more essential to the individual; but modern enterprise and invention have, through invasions upon his privacy, subjected him to mental pain and distress, far greater than could be inflicted by mere bodily injury.

It took some years for these privacy rights to be recognized by a state supreme court (Georgia was the first, in 1905), and some states have yet to do so (the New York Court of Appeals, for instance, rejected the first right to privacy claim, in 1903, and has never changed its mind). However, since "The Right to Privacy" was published, a number of state courts and legislatures have recognized one or more privacy torts. A number of states are also beginning to treat invasion of privacy as a criminal act, depending on the type of information that is disclosed.

1. Specific Offenses
There are four generally recognized types of privacy torts, all of which can far too easily be committed by children as well as adults:

1.   Misappropriation of an individual's likeness or image
2.   Intrusion upon seclusion
3.   Publication of private facts
4.   False light depiction

The following descriptions of these torts are not meant to be comprehensive, but to provide a basic overview of possible violations. Whether your child may have committed one or more of these torts will depend on the laws of the state in which you live. Be sure to consult a licensed attorney if you have questions or concerns about your child's potential liability.

The fifth category discussed below pertains to the growing number of state statutes that make invasion of privacy a criminal act. I expect that the number of states who add this new type of crime to their law books will steadily increase.

Misappropriation of an Individual's Likeness or Image or Right of Publicity

When Brandeis and Warren wrote their famous law review article, they were primarily concerned about what we would now consider misappropriation of image, i.e., the growing tendency of photographers and newspapers to use an individual's likeness for commercial purposes without consent. Samuel Warren in particular was an intensely private man who resented the photographers who were so enamored of his high-society wife, Mabel Bayard Warren, a member of an upper-crust Delaware family and daughter of a secretary of state. The Warrens also were close friends with President Grover Cleveland's young and attractive wife, Frances "Fanny" Cleveland, who was herself a popular target for 19th-century paparazzi. (President Cleveland was reportedly infuriated that Fanny's photo was used without permission to advertise a wide range of products.)

The tort of misappropriation today allows an individual to recover damages for the commercial use of a protected attribute (typically someone's name or image). In some cases, damages for noncommercial use also can be recovered if the defendant used the attribute for personal benefit. A misappropriation claim might involve the use of someone's name or identity on a blog or social networking site, or the use of someone's image to promote or endorse a website without permission.

A claim for misappropriation of a person's right of publicity is very similar, and the two claims are often confused. The distinction tends to depend on the level of the victim's fame. If, for instance, your child uses the image of a celebrity to promote a website or product, he or she could be guilty of not only misappropriating the celebrity's image, but also interfering with the celebrity's ability to capitalize on his or her fame. That claim is a more difficult one to prove for someone who is not particularly well known by the public.

Intrusion upon Seclusion

If your child has dreams of becoming an amateur paparazzo and starts climbing into backyards to take photos of classmates for his blog or website, he or she may be liable for the tort of intrusion upon seclusion. The gist of this claim is that the defendant has taken photos or videos of someone engaged in private activity in a private place, then made those images public.

This tort also applies when someone gains access to and publicizes an individual's private data. So, for instance, if your child somehow gains access to a classmate's e-mail account or text messages (perhaps by finding a phone in the hallway), then rebroadcasts those messages, a claim for intrusion upon seclusion could be filed.

Publication of Private Facts

The tort of publication of private facts is similar to the tort of intrusion upon seclusion, but refers to a specific class of information. In order for someone to make a claim for damages under this tort, four elements must be shown: (1) that there was a public disclosure of certain facts; (2) that those facts were private; (3) that disclosure of those facts would be offensive to a reasonable person; and (4) that the facts disclosed were not newsworthy, i.e., of legitimate public concern.

There are a wide variety of circumstances under which a child could face liability for publication of private facts. Imagine, for instance, that your daughter learns her classmate was absent from school for a week because she had an abortion, a fact which your daughter then posts to Facebook. Or imagine that your son tweets the undisclosed fact that the captain of the football team is gay. The fallout from such disclosures can be severe, and it's not difficult to imagine that substantial damages could result.

False Light Depiction

More recently, a new type of lawsuit has arisen: the claim of depiction in a false light. To a large extent, the claim of false depiction is similar to defamation. The primary difference is that, unlike defamation, truth is not a valid defense. In fact, the whole basis of such claims is that the publication by the defendant, while technically true, is phrased in such a way—or leaves out significant context—so that the reader is left with a false (and defamatory) impression of the plaintiff.

A good example is provided by a case that came before the Florida Supreme Court in 2008. In *Anderson v. Gannett Company, Inc.,* Joe Anderson filed a lawsuit against a Florida newspaper which mentioned, in an article about a paving company, that Anderson had "shot and killed his wife." Later in the article, the newspaper noted that the shooting was a hunting accident. Anderson argued that the manner in which the information was presented, even though technically true, created the false impression that he had intentionally shot his wife. The question before the Florida Supreme Court was whether to recognize a new type of claim—i.e., depiction in a false light—or to impose defamation's usual two-year statute of limitations, which would have barred Anderson's claim. In this particular case the court chose not to recognize the new claim, a position also taken by a number of other state courts.

2. The Scope of the Problem

So far, there have been relatively few civil lawsuits filed against students for violating the various privacy rights discussed above. There are a number of reasons why plaintiffs may be reluctant to sue students for invasion of privacy: for instance, there are other legal mechanisms for punishing inappropriate or invasive online statements (cyberbullying, harassment, etc.); issues of proof can be problematic, particularly in the relatively fluid environment of teen relationships; and teens are not typically good targets for damage claims. Even when a prosecutor can show that the parents are liable—a difficult task—the parents' pockets might not be particularly deep. It's also worth noting that it can be nearly as expensive to file an invasion of privacy lawsuit as it is to defend one, so without a reasonable prospect of recovering damages, it may not be worthwhile to initiate litigation.

The most notable criminal prosecution for invasion of privacy, as noted earlier, is the ongoing action against former Rutgers students Dharun Ravi and Molley Wei for allegedly recording images of Ravi's roommate, Tyler Clementi, engaged in a sexual encounter with another male. Clementi later committed suicide by jumping off the George Washington Bridge.

3. Consequences

The consequences for an invasion of privacy will depend on whether your child faces a civil lawsuit or criminal prosecution (in a worst-case scenario, like the Rutgers case, your child could face both types of actions).

Civil lawsuits are an effort to recover the damages that the plaintiff has suffered as a result of the wrongful action of the defendant. Such lawsuits typically claim compensation for the actual harm that the plaintiff suffered as a result of the invasion of privacy, as well as emotional distress associated with the invasion. In addition, if the invasion of privacy is shown to be willful (as opposed to merely false or reckless), the plaintiff may be awarded punitive or exemplary damages. Since punitive damages are not tied to a specific harm, but instead are intended to be an amount large enough to express the jury's disapproval and prevent future occurrences, the amount of damages awarded can be quite large.

The criminal penalties that can be assessed if your child is convicted of invasion of privacy vary from state to state. The New Jersey invasion of privacy statute (which is receiving the most attention right now), provides for a penalty of between three and five years in prison and/or a fine of up to $15,000. Distribution of privacy-invading videos or photos is a separate offense with identical penalties. As often happens after high-profile cases such as that of Tyler Clementi, New Jersey legislators are considering heightening the penalties. A bill introduced by Rep. Shirley Turner (D-NJ) in October 2010 would increase the possible prison time to between five and ten years, and increase the fine by a factor of ten, up to a maximum of $150,000. Even if Turner's bill does not pass, it is likely that the Clementi case will inspire other states to add invasion of privacy to their criminal statutes.

4. Investigation and Prevention

The tools for determining whether your child might be at risk for an invasion of privacy claim are essentially the same as those for defamation, so there's no need to repeat them here. The same preventative measures apply as well, with a couple important additions. The first is to inculcate in your children a respect for other people's privacy. Given that the legal right to privacy is still relatively new, it's amazing how deeply ingrained the idea has already become in American society.

Implicitly and explicitly, we teach our children the concept of privacy from an early age, and it's not uncommon to hear one sibling yelling at another, "Stay out of my private space!"

But kids do seem to have a harder time grasping the idea of privacy. Both the media and the Internet undercut the concept, for many of the same reasons Brandeis and Warren anticipated a century ago. In a world where the media peddles every detail that can be purchased or stolen about celebrity lives, and where millions of people—teens and adults alike—post endless details about their own lives on blogs or social networking sites, it can seem as if private space no longer exists. The law, however, insists otherwise, and rightly or wrongly your child can be sued for invading someone else's private sphere.

The second concept to teach your children, closely related to the first, is discretion. It may be that your child learns something accidentally, or is told something by a classmate in confidence. Again, the open nature of the Internet and the compulsive disclosure urged by social networking sites make it far too easy to spread a salacious bit of gossip. But your child does not need to be the Julian Assange of the eighth grade in order to earn the admiration and respect of his peers. An important part of each child's education is that sometimes the right thing to say is nothing at all.

Chapter Eight
Internet Addictions

In this chapter and the next, our focus shifts away from communication-based cybertraps, which tend to expose children chiefly to civil liability, and toward behaviors that are more explicitly criminal.

Online Gambling

1. The Crime

Given human nature, it's not terribly surprising that the first two highly successful Internet industries were adult entertainment and gambling. Congress has made several attempts to restrict adult content online, none particularly successful, but legislators have had more luck limiting online gambling.

The legality of online gambling has always been suspect because of a 1961 law, the Wire Act, which bars bookmakers from accepting bets over the telephone. The language of the Wire Act is very specific, however, and efforts to apply it to more modern technologies have been legally questionable. In an effort to limit the perceived harms of the growing online gambling industry, Congress in 2006 passed the Unlawful Internet Gambling Enforcement Act (UIGEA). While the law doesn't ban online gambling outright, it prohibits banks and other financial institutions from processing transactions with gambling sites.

Unsurprisingly, the gambling industry has enormous financial resources available for legislative lobbying. Within months of the law's passage, Congress went in the other direction and began considering legislation to legalize and regulate online gambling. In July 2010, a bill to do just that, sponsored by Rep. Barney Frank (D-MA), passed the House Financial Services Committee, but it was never taken up by the full House and thus died at the end of the 2010 session. The New Jersey Senate, meanwhile, in November 2010 became the first state legislative body to pass a bill to explicitly authorize online gambling. In early 2011 the measure was still awaiting consideration by the state assembly. Under the terms of the bill, only New Jersey residents would be permitted to gamble online. Whether the law will be reconciled with the UIGEA remains to be seen.

Currently, it's not a crime, per se, for your child to gamble online. The real cybertrap here is what a child has to do to access the funds necessary to gamble. The easiest source of funds is a credit card, which a surprising number of teens already have in their possession (15 percent of high school students had credit cards of their own as of 2008). In theory, U.S. credit card companies are barred by the UIGEA from processing transactions involving gambling sites, but that law has proven difficult to enforce, due to the sheer volume of websites that offer gaming.

Another way teens can fund their online gambling is by setting up electronic withdrawals from a bank account. Again, those transactions should, in theory, be blocked by the UIGEA, but many of the gambling sites offer their users detailed instructions for skirting the law by transferring funds from their bank to an intermediate financial company or electronic payment processor that then does business with the gaming site. These payment processors—like the gambling sites themselves—are usually set up offshore, which means they're not governed by the UIGEA. And there are so many that it's virtually impossible for American banks to keep track of which payment processors have close relationships with gambling sites.

If this all sounds too complicated for your child to figure out, think again. Kids are surprisingly adept at even complicated online activity, and the gaming sites themselves provide roadmaps to walk their users through these transactions.

It's bad enough if your child uses his or her own bank account to gamble, but presumably the amount of available funds is limited. It can become even more tragic if your child gets access to *your* banking information and sets up automatic withdrawals from your account. All the financial information your child would need to set up those transfers is likely in your files, or readily available on your home computer.

If your child uses a credit card to gamble online and, in the process, racks up a large debt, you may be able to get the credit card company to void the charges, since under the UIGEA those transactions are technically illegal. However, there are frequently time limits for challenging credit card charges, so it pays to be on top of your child's card use. If, on the other hand, your child has gambled with money drawn directly from a checking or savings account, it will probably be more difficult to recover.

2. The Scope of the Problem

Teen gambling is a growing problem, one that is exacerbated by state lotteries, cable television, and the Internet. Lotteries and most other types of gambling were once illegal in the United States, but over the last half century there's been an explosion in the number of venues for legal gambling. Forty-two states run one or more lottery games, and nineteen states now license casinos. The World Series of Poker, which has been around for some forty years, in 2003 became an absolute staple of cable television when ESPN expanded its coverage to include the entire tournament and not merely the final table. Gambling often features prominently (and favorably) in television shows and movies. And of course the Internet offers a seemingly endless array of gambling websites, most of which offer "free" games to lure people into spending their time (and eventually their money).

The statistics are frightening. According to one recent study by the Institute for Research on Gambling Disorders, over 70 percent of Americans between the ages of fourteen and nineteen have gambled in the past year. In a separate survey, University of California-Berkeley researchers concluded that nearly 20 percent of young men "gamble online regularly." They're no doubt encouraged by stories like that of Joe Cada, the 2009 World Series of Poker champion, who was just twenty-two when he claimed the $8.5 million top prize. Cada told ESPN that he honed his skills playing Internet poker, racking up as many as two thousand hands per day (which works out to nearly three-quarters of a million hands of poker in a given year).

Unfortunately, very few teenage gamblers will experience the kind of success that has earned Cada free rooms in Vegas casinos and a security escort. For the vast majority of teens, gambling in general and online gambling in particular will provide a crash course in the law of averages, unfavorably distorted by the typical house cut and software run by offshore companies that may not always play fairly.

Researchers and social workers have noted a steady rise in the number of teens with large gambling debts, and a corresponding rise in criminal activity associated with efforts to get more money to gamble or pay off debts. Teens addicted to gambling, particularly online gambling, frequently cut themselves off from friends, quit extracurriculars, and even drop out of school. An estimated 3 to 5 percent of teen gamblers will grow up to be chronic, problem gamblers.

3. A Cautionary Tale

During the course of my research, I came across a compelling story that described the painful consequences of teen gambling. The article—originally posted on the freelance blogging site Info Barrel—tells the story of a single mother, Leslie Cramer, living in California and raising fifteen-year-old Andrew, the youngest of her five children. Like most Americans, Leslie was struggling in a down economy; she had recently lost a private nursing gig, and was having a difficult time replacing the income from that job. While she searched for a new position, she and Andrew were living off her savings.

In the fall of 2009, his sophomore year in high school, Andrew was excelling. He had excellent grades, was a member of the high school football team, and had developed extensive skills in web design and video production. While his mother worked and looked for better-paying jobs, Andrew spent a lot of time hanging out with his older brother, who made one critical error: he introduced Andrew to online poker.

It wasn't long before Leslie began noticing changes in Andrew. He quit the football team, started bringing home lower grades, and stopped spending his time on web development. Leslie knew he was spending a lot of time playing online poker, but both he and his brother assured her that it was just for fun and no money was involved. Overwhelmed by work and financial pressures, Leslie did not pursue the matter, thinking the changes in Andrew were just the typical symptoms of teen development. In fact, she was pleased that her son had found an interest he could share with his older brother.

Over the next several weeks, Leslie began noticing sporadic decreases in the balance of her savings account. She initially wrote these off as bills she'd forgotten, but when her balance continued to shrink, falling rapidly over the course of a few days, she finally contacted her bank. The bank informed her that over $7,000 in online transactions had been processed over the previous few months. Further investigation revealed that the money had been paid to PokerStars.com, an online gambling site based on the United Kingdom's Isle of Man.

Leslie confronted Andrew, who eventually admitted that he was responsible for the lost money. He was worried about their financial situation, he said, and about how hard she was working. He thought he could help by winning at online poker, so he got access to her banking information and used it to process electronic checks into a PokerStars account.

The PokerStars site acknowledges that "a small percentage of players will attempt to play when they're underage," and so it has instituted age verification. However that seems easy enough for teenagers to skirt. PokerStars requires players to click a box acknowledging they're eighteen, and when a player creates an account the site says it "collect[s] their name, address, and birth date to confirm that the player is at least eighteen years old." PokerStars also claims that it "does not target underage players with [its] marketing and advertising." It's worth pointing out, however, that ever since Chris Moneymaker (and yes, that is his real name) won the World Series of Poker in 2003 after winning a qualifying tournament on PokerStars, the site has been an enthusiastic sponsor of the tournament and a heavy advertiser during the tournament's television broadcasts on ESPN. It is safe to assume that hundreds of thousands of teens have seen ads for PokerStars.com.

Obviously, a kid who is intent on playing online poker is not going to be deterred by the need to lie about his or her age. Nor is the funding issue difficult: if a child can get access to the necessary financial information, PokerStars and other sites helpfully provide all the instructions needed to play with a credit card or set up electronic transfers from a bank account.

There are, however, some unanswered questions in this story that other parents should be asking themselves. How easy, for instance, was it for Andrew to get access to his mother's banking information? What additional questions should Leslie have asked when she learned her sons were playing poker online? Are the bank and PokerStars.com liable in any way for the losses, since the bank should not have processed the withdrawals, and PokerStars.com should not have allowed a minor to play? Or is Andrew's own conduct—lying about his age and misappropriating his mother's money—a sufficient defense for both?

In the short run, Andrew's online gambling, however well meaning, created more intense financial difficulties for his mother, and damaged family relationships. Under different circumstances, he also could have faced grand larceny charges. This tale, fortunately, had a happy ending: the bank refunded the money that Andrew paid to PokerStars. com. But it's a sobering story that easily could have ended much less positively—and I appreciate the willingness of Leslie and her daughter, freelance writer Emily Murray, to share it with others.

4. Consequences

Currently, there is no federal law that specifically prohibits individuals (even teens) from gambling online, although nine states have adopted statutes that make it a crime (typically a misdemeanor) to do so. A survey of sites that cover online gambling law makes it clear that the likelihood of anyone actually being prosecuted for online gambling is quite low.

If your child does gamble online (and particularly if he or she develops a gambling addition), the consequences are more likely to stem from the obsession itself and the actions necessary to fund the activity. Online gambling is sometimes referred to as the "crack cocaine" of the gaming industry, an activity that produces a measurable (and addictive) chemical reaction in the brain. The speed with which online gaming takes place adds to the chemical hook. It's not surprising that parents report that, as the obsession grows, gambling teens lose interest in activities, schoolwork, friends, and even family. While your child may not face legal consequences, the impact is no less severe: poor grades, diminished college and job prospects, and a blow to mental and physical well-being.

Criminal consequences may also be in the offing, of course, depending on how your child funds his or her online gambling activity. If funds are misappropriated from your bank account or credit card, or if money is stolen from other sources to pay off gambling debts, additional and more serious crimes are being committed, and those crimes certainly could be prosecuted.

5. Investigative Tools
The first question you need to answer is whether you care that your child might be gambling online. The Minnesota Council on Compulsive Gambling put that question to parents with children under the age of fifteen, and fewer than one in ten said they would be concerned if they found out their child was gambling online. Given the potentially serious financial and emotional consequences, that strikes me as a shockingly low number.

But let's assume for the moment that you're in that 10 percent who *would* be concerned. The next question is, how can you find out whether your child *is* gambling online? The answer involves talking with your child, and paying close enough attention to his or her behavior to observe the potential warning signs:

- Frequent requests to borrow money from you, other family, or friends
- The sudden appearance of gambling ephemera (scratch tickets, betting slips, etc.)
- Sudden windfalls of cash
- Reduced interest in extracurricular activities
- Behavioral changes and sudden mood swings
- A sudden or disproportionate interest in sporting events, or a surprisingly intense reaction to the outcomes of games
- An extensive amount of time spent on Internet gambling sites, supposedly just for "fun"
- Money missing from your home or bank account
- Unexplained charges on your child's credit card or your own

As most parents will agree, some of these warning signs point to nothing more serious than the onset of adolescence. But if you observe several of them around the same time, particularly those involving money, it's time to get more focused about looking into your child's online behavior.

6. Prevention

If you know or even suspect that your child has an interest in online gambling, there are a number of simple steps you can take to minimize the potential damage.

The first and most critical: talk to your child about the dangers of gambling and its addictive qualities. The amount of pro-gambling content kids see in the media is staggering, and it will take multiple conversations about the dangers of gambling to effectively counter it. The more time you can invest in talking with your kids about what they're doing online and what the risks are, the less likely you'll find yourself sitting nervously in your bank manager's office and wondering where your money went.

A second step is to consider a requirement that your child's computer be used in a common space, where you can visually observe what's taking place. Of course that's not a perfect solution. It may be physically difficult to arrange, it runs counter to teaching children to work in quiet spaces without distractions, and you may not be home enough to make it a useful requirement. Additionally, the majority of households have two working parents, which tends to leave teens with a lot of unsupervised time, whether in their own homes or those of friends. Even when adults are present, teens are amazingly adept at rapidly switching screen views on a computer to hide undesirable activity.

Given all those factors, electronic supervision may be the only viable alternative for determining whether your child is gambling online. Appendix A lists a variety of different software packages that can help you do so effectively. Some parents may balk at the idea of tracking everything their children do online, but depending on the age and maturity of your child, it may be the only responsible choice. For children who protest vehemently, here are a couple useful responses.

First, until your child reaches the age of eighteen, privacy is a privilege, not a right. The more maturity a child can demonstrate, and the more confident you are that he or she isn't getting into trouble online, the more privacy you'll honor.

Second, it's worth pointing out to kids that even adults tolerate a certain amount of surveillance—speed traps, for instance—as a necessary aspect of using a potentially dangerous instrument like a car. Since most kids operate computers and mobile devices long before we would trust them behind the wheel, a certain amount of surveillance is a reasonable step.

Third, you should take basic but thorough steps to protect your financial information and access to your computer. A locked cabinet or safe for financial records is a good idea, as is a shredder for disposing of confidential records you no longer need (which is also a recommended means of preventing identity theft by people outside the household). Every computer can handle multiple user accounts, so you should set one up for yourself with a strong password. Keep in mind, of course, that those efforts will be wasted if you hang the key to the file cabinet on your computer monitor, or leave behind a Post-It note with your password in plain view.

Lastly, it's important to stay on top your own financial affairs. If you see a bank balance that looks odd, strange debits posted to your account, or unexplained and cryptic credit-card charges, look into the issue immediately. The sooner you act, the better. Keep in mind that most credit cards have time limits for challenging charges, so if you let things slide too long, it can be that much more difficult to get charges reversed and recover your money.

Online Tobacco, Alcohol, and Drug Purchases

When I began writing this book, it hadn't occurred to me that tobacco, alcohol, or drugs would be among the cybertraps that could threaten kids. Of course I was aware that all three can pose risks for children, but I didn't think the Internet would be much of a factor. I was saddened to discover how wrong I was. As a matter of fact, there is an active—and growing—online market for all kinds of illegal substances, and it's all too easy for kids to tap into it.

1. The Crime
Until 1984, the drinking age in the United States varied from state to state. That year, Congress passed the National Minimum Drinking Age Act, which stipulated that 10 percent of federal highway funds would be withheld from any state that refused to raise the minimum drinking age to twenty-one. With some grumbling (particularly from Texas), all the states eventually complied. It is now illegal for individuals under the age of twenty-one to purchase alcohol or to consume it in public locations. It is also illegal for individuals over the age of twenty-one to purchase alcohol for minors.

There are, of course, both federal and state laws against the use of various controlled substances and the misuse of prescription medications. Legal penalties attach to the possession of controlled substances, as well as their production, sale, distribution, use, and even their advertising. That is to say little of the penalties for operating a motor vehicle under the influence of drugs. Teens who wind up with any kind of drug conviction on their record will find it difficult to get into college, get student loans, find a job, or enter the military. If the offense is severe enough, it may even result in a prison sentence.

2. The Scope of the Problem
Even without the aid of the Internet, underage drinking and drug use are enormous problems in the United States. Let's start with alcohol, which according to the Centers for Disease Control and Prevention is "the most commonly used and abused drug among youth in the United States, more than tobacco and illicit drugs." According to the CDC, 11 percent of all the alcohol consumed in the United States is drunk by people between the ages of twelve and twenty.

Even more disturbing are the findings of the 2009 Youth Risk Behavior Study, which asked high school students about their alcohol consumption during the previous thirty days:

- 72 percent said they drank at least some amount of alcohol
- 24 percent admitted to binge drinking (which the CDC defines as approximately five drinks in two hours for males and four drinks in two hours for females)
- 10 percent said they'd driven after drinking alcohol
- 28 percent said they'd ridden with a driver who had been drinking

The general statistics about tobacco and drug use are no more encouraging. The CDC reports that more than 80 percent of current adult smokers began smoking before they turned eighteen. Every single day, approximately a thousand people under the age of eighteen begin smoking on a daily basis. As of last year, the CDC reported, roughly 17 percent of daily cigarette smokers were high school students, and about 5 percent were in middle school. The National Institute on Drug Abuse is somewhat more encouraging, finding that fewer than 3 percent of eighth graders are daily smokers (though one in five have at least tried cigarettes) while the daily smoking rate among high school seniors is just over 11 percent (with just under half having tried cigarettes at some point).

As for drugs, an annual survey conducted by a division of the Office of National Drug Control Policy found that illegal drug use rose by 9 percent in 2009, the highest rate of increase in the past decade. Much of the increase was driven by a sharp rise in the use of Ecstasy (MDMA), up thirty-seven percent from 2008, and methamphetamines, up sixty percent. Overall, 8.7 percent of Americans twelve and older reported using illegal drugs in 2009.

Since 1975, the National Institute on Drug Abuse has conducted a detailed study called Monitoring the Future, in which it has tracked teen use of various illegal substances. Here are some of the significant findings for 2009:

Monitoring the Future Study:
Trends in Prevalence of Various Drugs for 8th-Graders, 10th-Graders, and 12th-Graders, 2007-2010 (in percent)*

	8th-Graders				10th-Graders				12th-Graders			
	2007	2008	2009	2010	2007	2008	2009	2010	2007	2008	2009	2010
Any Illicit Drug Use												
Lifetime	19.0	19.6	19.9	21.4	35.6	34.1	36.0	37.0	46.8	47.4	46.7	48.2
Past Year	13.2	14.1	14.5	[16.0]	28.1	26.9	29.4	30.2	35.9	36.6	36.5	38.3
Past Month	7.4	7.6	8.1	[9.5]	16.9	15.8	17.8	18.5	21.9	22.3	23.3	23.8
Marijuana/Hashish												
Lifetime	14.2	14.6	15.7	17.3	31.0	29.9	32.3	33.4	41.8	42.6	42.0	43.8
Past Year	10.3	10.9	11.8	[13.7]	24.6	23.9	26.7	27.5	31.7	32.4	32.8	34.8
Past Month	5.7	5.8	6.5	[8.0]	14.2	13.8	15.9	16.7	18.8	19.4	20.6	21.4
Daily	0.8	0.9	1.0	[1.2]	2.8	2.7	2.8	[3.3]	5.1	5.4	5.2	[6.1]
Alcohol												
Lifetime	38.9	38.9	[36.6]	35.8	61.7	[58.3]	59.1	58.2	72.2	71.9	72.3	71.0
Past Year	31.8	32.1	30.3	29.3	56.3	[52.5]	52.8	52.1	66.4	65.5	66.2	65.2
Past Month	15.9	15.9	14.9	13.8	33.4	[28.8]	30.4	28.9	44.4	43.1	43.5	[41.2]
Daily	0.6	0.7	[0.5]	0.5	1.4	[1.0]	1.1	1.1	3.1	2.8	2.5	2.7
Cigarettes (any use)												
Lifetime	22.1	20.5	20.1	20.0	34.6	[31.7]	32.7	33.0	46.2	44.7	43.6	42.2
Past Month	7.1	6.8	6.5	7.1	14.0	[12.3]	13.1	13.6	21.6	20.4	20.1	19.2
Daily	3.0	3.1	2.7	2.9	7.2	[5.9]	6.3	6.6	12.3	11.4	11.2	10.7
1/2-pack+/day	1.1	1.2	1.0	0.9	2.7	[2.0]	2.4	2.4	5.7	5.4	5.0	4.7
MDMA												
Lifetime	2.3	2.4	2.2	[3.3]	5.2	4.3	5.5	6.4	6.5	6.2	6.5	7.3
Past Year	1.5	1.7	1.3	[2.4]	3.5	2.9	3.7	[4.7]	4.5	4.3	4.3	4.5
Past Month	0.6	0.8	0.6	[1.1]	1.2	1.1	1.3	[1.9]	1.6	1.8	1.8	1.4
Methamphetamine												
Lifetime	1.8	2.3	[1.6]	1.8	2.8	2.4	2.8	2.5	[3.0]	2.8	2.4	2.3
Past Year	1.1	1.2	1.0	1.2	1.6	1.5	1.6	1.6	[1.7]	1.2	1.2	1.0
Past Month	0.6	0.7	0.5	0.7	0.4	[0.7]	0.6	0.7	0.6	0.6	0.5	0.5

Data in brackets indicate a statistically significant change from the previous year

Courtesy of the National Institute of Drug Abuse.
Available at http://www.drugabuse.gov/infofacts/HSYouthtrends.html

3. The Role of the Internet and Mobile Devices
The common consensus right now is that in all three cases—tobacco, alcohol, and drugs—online purchasing by teens is still a relatively new phenomenon, but a steadily growing one.

As early as 2002, the Minnesota Department of Health found that websites selling tobacco did a poor job with age verification, packaging, and delivery restrictions. For instance, every single company examined by Minnesota shipped cigarettes in unmarked packages without return addresses. In one instance, a package was delivered to an eight-year-old child. The study also found a significant loss of tax revenue, since none of the cigarette packs carried state tax stamps.

Good Morning America did a segment in March 2001 in which it found that kids in Salt Lake City as young as ten were able to order cigarettes online and have them shipped directly to their homes. A sting operation operated by the Utah Attorney General's Office found that more than half of the orders placed by children under the age of eighteen were filled by online retailers, no questions asked. In New York, *GMA* found, an eight-year-old child was able to successfully order three cartoons of Kool cigarettes.

When it comes to alcohol, the story is a similar one. Nearly five years ago, Teen Research Unlimited—responding to a request by the Wine & Spirits Wholesalers of America—surveyed teens about online alcohol purchases. About 2 percent of the roughly half-million teens surveyed admitted to buying alcohol online, and another 12 percent knew of at least one friend who had done so. That appears to be the last comprehensive survey taken on the issue, but a review of online message boards leaves little doubt that kids are turning to the Internet when they're unable to get alcohol from their older siblings, friends, or parents. Much like websites that sell cigarettes on the Internet, the mechanisms used to check the age of online alcohol consumers are either weak (simple check boxes) or non-existent.

Ironically, it may be even easier for kids to get drugs online than either alcohol or tobacco. In a 2006 study, the Partnership for a Drug Free America found that 20 percent of teens were abusing prescription drugs, many of which—Xanax, Valium,

Vicodin, and Ritalin—can easily be purchased online. A separate Columbia University study found that 94 percent of websites selling prescription drugs did not require a prescription before completing a sale (and in fact, many kids reported that getting prescription drugs was easier than purchasing beer). More recently, kids have taken to making YouTube videos demonstrating the effects of taking various herbs or snorting crushed bath salts, substances which are currently not regulated by the Food and Drug Administration or the Drug Enforcement Agency.

Just last month, three Canadian teens were hospitalized after ingesting small quantities of psychoactive drugs they had purchased online. Police declined to reveal the name of the substance out of concern for copycat incidents, but as a *Toronto Sun* reporter pointed out, a host of possibilities can be located online with minimal effort.

Of course the Internet is not the only electronic tool that can be used by kids to facilitate their use of illegal substances. A number of compelling stories are recorded on the Office of National Drug Control Policy's anti-drug website (http://www. theantidrug.com/), and they offer chilling insight into the power of electronic technology to promote addiction. Here are just a few excerpts:

> **Amy's Story**: "So how did I become so addicted? It was actually very easy, thanks to my cell phone. Yes, in 8th grade I got a cell phone . . .I kept all my drug dealers close by. In fact, I could press a couple of buttons on my cell phone, and there they were—in my address book. Whether I was at school or on vacation with my family, I always had a dealer within 10 minutes from me. With a press of a 'detail' button for each contact, I had all the information I needed—what types of drugs they sold, where they lived, and how to get a hold of them. Normally, I'd call or text message a dealer around 2:00 during the school day and, by the end of classes, I was hooked up."

Cheyenne's Story: "At the time (age 12), I was visiting a lot of chat rooms, meeting druggie guys, and they would [instant message] me asking if I drank or got high. I didn't think too much about giving them my cell phone number. We would end up meeting to drink, get high or mess around with prescription drugs . . .When I wasn't hanging out in chat rooms, I was posting stuff to my MySpace profile. I lied about my age and posted pretty seductive pictures to attract guys. Most were between 16-25 years old. Of course, I attracted the 'druggie' types because my profile page was filled with talk about smoking and drinking. I dug the attention and acceptance! But the drugs really took a toll. I was rushed to the hospital three times for drug and alcohol abuse."

Sean's Story: "It was when the [drug] screening began that I started going to the Internet to find ways to pass a drug test. Guess what? If you type in 'how to beat a drug test' into Google, you can get a ton of websites! From that point on, the Internet became my first source of drug information. I found out about 'smoking herbs,' how to grow marijuana, make crack, recipes for LSD, how to make pot more potent, you name it—I found it! I stumbled onto a site called Erowid. com through a Google search. It was a mecca for pro-drug information! . . . Eventually, I linked my IM to the site so my friends could access the information, too. We would order drugs online, even prescription drugs. All we needed was a credit card or billing address. Pretty easy."

4. Investigation and Prevention

What's not easy, by any stretch of the imagination, is monitoring and supervising the myriad ways your kids can get hooked on addictive activities or substances. The Internet can often look like nothing more than an addiction-delivery mechanism, and it's enormously tempting to think that turning it off is the best solution. But really, it isn't—the Internet is such a fundamental part of our society that you'd only be handicapping your children's development. Besides, your kids would likely find surreptitious ways to go online anyway, against your wishes. The better approach is to engage your children on a regular basis, place reasonable limits on their Internet activity, and do slightly more active surveillance than you think they actually need.

As corny as public service campaigns sometimes sound ("Parents! The Anti-Drug"), you *are* the first and foremost tool in fighting addiction. It's not enough to provide a comfortable home in a nice neighborhood, plenty of academic and extra-curricular resources, or even a relatively stable family life—plenty of kids with those advantages have developed serious, even life-threatening addictions. What's needed from an early age are positive engagement and clear boundaries. Teens need to know what your expectations are for their behavior, both online and off, and the consequences for violating those rules. Do some additional research about these issues yourself, so you can have informed discussions with your children. Listen carefully to their concerns and let them know what worries you.

The consumption of tobacco, alcohol, or drugs, more so than online gambling, tends to be social. Talk with your children. Get a sense of who their friends are, and talk to those friends' parents about how parties and other get-togethers are supervised, what activities are permitted, and so on. Keep an eye on the electronic conversations your children have with their friends, and don't be reluctant to ask some questions about names you don't recognize. Learn how to check the contact list of your child's phone, and ask about any entries that don't make sense.

In the online realm, make sure you visit your child's social networking pages. If your fourteen-year-old daughter is posting provocative pictures on her MySpace page as a way of attracting older men with drug connections, that's obviously something to investigate. If there's a lot of drug lingo or slang on your child's Facebook wall, or chatter about last weekend's alcohol-fueled blowout at a parentless house, there are conversations that need to be had. Inevitably, this also raises the specter of electronic surveillance. It may be that the only way to keep track of what your child is doing online is to enlist the help of one of the surveillance suites discussed in appendix A.

Lastly, spend the time necessary to network with friends, neighbors, teachers, and school administrators. Enlist their help in setting boundaries for your children and educating them about the physical and emotional risks associated with tobacco, alcohol, and drugs. Most schools already include information in their curricula about these dangers, but don't leave your child's well-being solely in someone else's hands. Educate yourself about what is being taught in school, and question whether your kids are being taught early enough and thoroughly enough. Make it clear to your children that not only do you understand and support what the schools are teaching about addiction, but that it's merely a starting place for your concerns and your own family's boundaries.

5. Consequences

It's not called "The War on Drugs" for nothing—over the last thirty years, federal and state legislatures have adopted increasingly strict laws and severe penalties for individuals who possess, purchase, and sell illegal drugs, which includes prescription medicines without the appropriate prescription.

Unlike gambling laws, which are chiefly aimed at the businesses that run gambling operations or process payments for them, anti-drug laws are very much aimed at individual users and dealers. If your child purchases illegal or unauthorized prescription medicines online, he or she is committing a serious crime, one that is punishable under either federal or state law. The specific consequences depend on a variety of factors, including the specific drug involved, the quantity purchased, and what was done with it (use or distribution). The penalties cover a wide range, from probation and community service to lengthy incarceration and significant fines. Needless to say, a criminal conviction for illegal drug purchases or distribution will have life-long ripple effects.

It is important to note that even if your child purchases a substance online that is not currently illegal (such as methadrone), he or she could face serious civil liability for causing injury with it. For instance, if your child's friend gets methadrone or some similar substance from your child, and then suffers an injury or even death while under its influence, that child or his parents could sue your child (and possibly you as well) for the damages suffered.

After thirty years of the War on Drugs, the social and professional consequences of drug abuse should be crystal clear, but they're worth repeating: declining school performance, loss of scholarships and college acceptances, inability to get a professional license for various careers, not to mention the numerous health effects of ingesting questionable substances.

Chapter Nine
Identity Theft, Computer Fraud, and Hacking

"It's a young person's game." That phrase gets applied to many different vocations and avocations—sports, politics, military aviation, restaurant cooking, even magic—but nowhere is it more applicable than in the systematic misuse of computers for identity theft, fraud, and hacking. Painstakingly searching operating systems and programs for exploitable flaws requires a nimble mind, the ability to go long stretches without sleep, seemingly endless stores of energy, great focus, and freedom from most normal responsibilities—all qualities more likely to be found in the average thirteen-year-old than in those of us in what we hopefully refer to as middle age.

Computer technology (and technology in general) has upended the traditional hierarchy of professional knowledge. Not long ago, true expertise and knowledge could only be gained after years or even decades of effort and study (think law, history, languages, philosophy, etc.). But technology changes so fast that knowledge gained just a few years ago is rendered virtually useless by new software and hardware. In the early 1980s, I learned BASIC in high school, in part because my teachers thought everyone would need to know some programming language in order to be productive in the future. I might as well have studied Sanskrit. (An amusing exception to the rule of fast and disposable knowledge occurred as the year 2000 approached, when an army of gray-haired programmers skilled in the C programming language were brought out of retirement to fix the Millennium Bug they had inadvertently created.)

Given enough time and insufficient supervision, children can develop staggering levels of expertise. Consider the case of Jonathan James, who at sixteen became the first minor to receive a federal sentence for hacking (six months house arrest) when he was convicted in 2000. Among other things, he figured out how to crack into NASA and Department of Defense computers, snagging a copy of the international space station's environmental control program as a trophy.

Shortly after his trial, he gave an interview to PBS's *Frontline*, and perhaps without intending to do so illustrated the precise problem that makes so much teen online misbehavior possible: "[The government lacks] some serious computer security," James said, "and the hard part is learning it. I know Unix and C like the back of my hand, because I studied all these books, and I was on the computer for so long. But the hard part isn't getting in. It's learning to know what it is that you're doing."

Children become proficient hackers or identity thieves only when they have hundreds of hours of unsupervised time on the computer to develop their skills. There's an innate tension that develops: parents are often proud of their child's growing skills but give too little thought to how those skills might be misused. In James' case, his parents did try to rein in his computer use (he briefly ran away at thirteen), but clearly they had no idea that he was hacking into government computers until his arrest at sixteen. Four years later, the U.S. Secret Service began investigating him for possible involvement in a massive credit card theft from TJX (the parent company of T.J. Maxx). James committed suicide, after leaving a lengthy note defending his innocence.

Identity Theft

1. The Crime

The definition of identity theft is straightforward: the unauthorized use of someone else's personally identifying information—name, Social Security number, credit card number—to commit fraud or other crimes. Over the past several years, identity theft has become one of the fastest-rising computer-related crimes, in large part because our financial lives are so closely tied to our inherently insecure Social Security numbers. It doesn't help that we store so much of our personally identifying information—financial and otherwise— in electronic devices that are far too easily lost, stolen, or hacked.

As the Federal Trade Commission points out in its detailed online account, "Fight Back Against Identity Theft," victims can be left penniless from emptied bank accounts and fraudulent credit card charges, not to mention the potential negative effects to their credit scores. Beyond the financial devastation, falling victim to identity theft can also leave someone with a devastating sense of personal invasion and a loss of control.

How do identity thieves obtain the information they need? There are a variety of possibilities, only some of which require much technical skill, and almost all of which could be accomplished by a determined teen:

- **Dumpster Diving**: A surprising amount of confidential information is simply thrown out in the trash—bank statements, credit card offers, insurance forms, and more. Enterprising individuals will ferret through trash bags and dumpsters looking for that data, which can then be used to commit identity theft.
- **Change of Address**: It is disturbingly simple to obtain confidential information by filling out a change of address form and diverting an individual's mail to a new address, which saves all the mess of going through the trash.
- **Stealing**: Another way to obtain confidential information from the mail is to steal it directly from a mailbox or post-office box.

- **Skimming**: Thieves with access to a credit card can copy the information—as simply as using a copier or as sophisticated as an electronic card reader hidden in a shoe—to steal account information. It's most often committed by dishonest employees at a bar or restaurant, but also can be performed through elaborate means such as a fake ATM machine.
- **Phishing**: Scammers will often try to solicit confidential information with fake e-mails or pop-up ads that pretend to be from a financial institution or credit card company.
- **Pretexting** (also known as "social engineering"): An identity thief will pretend to be someone victims are likely to trust, like a bank employee or credit card representative, and ask to "confirm" personal information like account numbers or passwords.
- **Illegal Access**: With enough know-how, computers or networks can be broken into, allowing a perpetrator to collect personal identifying information.
- **Malware**: Online con artists often write and distribute software (viruses, Trojan horses, worms) that captures victims' financial records, account passwords, and other pertinent identifying information and e-mails that data back to the perpetrator.

In the vast majority of identity-theft cases, what's ultimately stolen is money, usually by employing the stolen personal information to apply for credit cards or other lines of credit. But it's important to remember that identity theft can apply to *any* personal information, not just financial information. It is increasingly common, for instance, for identity-theft charges to be filed for conduct related to cyberbullying.

If, for instance, your child were to gain access to another child's Facebook or e-mail account, that would be a form of identity theft. Similarly, if he or she were to impersonate another child and set up a harassing website—with the hope that the victim would be taunted or punished because of it (bullying by proxy)—that too could be prosecuted as identity theft.

2. The Scope of the Problem

It is difficult, if not impossible, to determine how many individuals under the age of eighteen have been prosecuted for (or convicted of) identity theft. The best source of information, the Juvenile Justice Bulletin produced by the Department of Justice, does not delineate "identity theft" as a unique category. Undoubtedly, some identity-theft arrests are included in the 2008 statistics for fraud (7,600 juvenile arrests), larceny-theft (324,100 juvenile arrests), and "all other offenses—except traffic" (363,000 arrests). As computer crimes become more common, both among adults and children, it's likely these types of reports will include more detailed statistics.

In the meantime, there are enough well-publicized examples of kids falling victim to identity theft that we know it's a very real problem. In an increasingly common type of case, three high school students in Newburyport, Massachusetts, created a Facebook page in another classmate's name and filled it with comments that were critical of other students. According to published reports, the victim was completely unaware of the fake page until his classmates began harassing him for what he'd supposedly written about them (a classic example of bullying by proxy). After several months of investigation, in February 2010, the three youths were identified and charged with identity theft (at the time, the Massachusetts cyberbullying statute was still being written; today they would likely also be charged with cyberbullying).

In another case, this one in the summer of 2009, Matthew C. Beighey—a high school student in Clifton Park, New York—was charged with third-degree identity theft after he came up with a rather novel approach for dealing with bad grades: locking his teachers out of the school's electronic report card system by inputting erroneous passwords three times into each of their accounts. This was actually Beighey's second identity theft incident; the year before, he'd been charged with a felony for using another student's ID to log into the school's computer system.

In a particularly disturbing example of teen initiative, eighteen-year-old Shiva Sharma, in January 2001, was arrested and charged with identity theft, grand larceny, and falsifying business records. Officials alleged that Sharma used a software program to harvest the e-mail addresses of thousands of AOL subscribers. He then created a fake AOL Web page and e-mailed his targets, telling them their billing information had been lost and directing them to the fake page to "re-enter" it. Sharma was able to capture credit-card information with a total credit limit of over $500,000, and proceeded to purchase $10,000 worth of electronic equipment and racing-car parts. Sharma had the dubious honor of being the first person charged under New York's identity theft law. After his third conviction at the age of twenty—for which he was handed a two-to-four-year prison sentence—Sharma was the subject of a lengthy profile in the *New York Times,* which included an illustration of the fake AOL page he created as a high school senior. The story also included a photo of Sharma's graduation photo—from Rikers Island High School, the educational program run by New York State at Rikers Island Correctional Facility.

3. Consequences

In 1998, identity theft became a federal crime when Congress passed the Identity Theft and Assumption Deterrence Act. That law allows for federal prosecution if someone "knowingly transfers or uses, without lawful authority, a means of identification of another person with the intent to commit, or to aid or abet, any unlawful activity that constitutes a violation of Federal law, or that constitutes a felony under any applicable State or local law" (18 U.S.C. § 1341).

In addition, as the Department of Justice notes on its website, many of the activities that constitute identity theft may also violate other federal laws. These include credit card fraud (18 U.S.C. § 1029), computer fraud (18 U.S.C. § 1030), mail fraud (18 U.S.C. § 1341), wire fraud (18 U.S.C. § 1343), and financial-institution fraud (18 U.S.C. § 1344). If your child is convicted of one or more of these crimes, he or she can face up to thirty years' imprisonment, hefty fines, and the possibility of criminal forfeiture of any "instrumentalities of the crime," including computers, related equipment, and anything purchased with proceeds from the criminal activity.

There's also the potential for state criminal charges. Every single state has adopted a law prohibiting identity theft or impersonation. If your child is facing state identity-theft charges, you should consult an attorney who practices in your particular state to advise you and your child on the relevant statutes.

4. Investigation and Prevention

Shiva Sharma's success as an identity thief (at least until he was caught) is a testament to not only his considerable programming skills, but also to a complete lack of parental supervision and the lure of the Internet's dark side. The *New York Times* article chronicles at length Sharma's progression from downloading pirated music to hanging out in Internet chat rooms, sharing bootlegged software, and eventually downloading tools to launch a hacking and identity-theft business. Bear in mind Sharma was able to do all this from a computer in the basement of his family home, and when packages of expensive equipment began showing up at the house with some frequency, there was apparently no effort by his parents to determine what was going on (a point that the *Times* story glosses over).

Sharma's exploits may seem a bit extreme—and they are—but his story is also an excellent primer on the warning signs that should trigger suspicion about what your child is up to on his or her computer:

- Your child is spending hours online in a location you can't easily observe.
- Your child suddenly seems to have large amounts of disposable income despite no apparent employment.
- Your child starts receiving packages containing expensive electronics, like stereo or camera equipment.
- Your child is increasingly secretive about what he or she is doing on the computer.
- Your child talks about getting back at, settling scores with, or otherwise bullying classmates, teachers, or administrators.

The sooner you can respond to clues like these, the sooner you can prevent or minimize the harm your child might be causing through identity theft or impersonation—and the consequences he or she could face for those very serious crimes.

The first step to take is to make sure you protect your own identifying information from possible misuse by your child. According to the Department of Justice, 13 percent of all identity theft is so-called "friendly theft," *i.e.*, committed by a family member or trusted friend. It's important, therefore, to make sure your financial information is secure and not easily obtainable. If there are documents you need to keep, make sure they're in locked storage; if there are documents you don't need, buy an inexpensive shredder at an office-supply store and shred anything that contains identifying information.

As with so many other cybertraps, preventing identity theft begins with an early grounding in ethics and cyberethics. Children need to be educated as early as possible about the boundaries of proper behavior, both online and off, and to appreciate the tremendous damage they can do by misusing other people's information. It's also useful—

as with other cybertraps—to periodically check in on how your children are using their electronic devices. Many of the low-level instances of identity theft—fake Facebook pages, e-mail account hacks, etc.—can be discovered by examining your child's computer (particularly the browser history and records of communication), or by hiring a local computer technician to assist you.

Realistically, if your child has advanced computer skills and has been spending time in hacker chat rooms or on hacking websites, he or she may also have the skills to block the installation of surveillance software, or to detect and remove it. Needless to say, that's a pretty good indication that you should get some outside assistance to help you figure out what's being done with the computer. In most cases, a local computer technician can give you a pretty good idea of what's being done with the computer, but it is more likely that you'll need the services of a computer forensics expert, who can usually recover much more evidence of the computer's recent activity.

Computer Trespass: Hacking and Cracking

1. The Crime

For those of us with long histories of computer use, there's a certain irritation that the word "hacking" has been transformed by careless legislators and journalists into a pejorative catch-all for computer crimes. The term's original use—in the late 1950s and early 1960s—encompassed an attitude towards systems (mechanical, electrical, pedagogical, political) that was highly inquisitive, fundamentally democratic, and more than a little rebellious, but interestingly *not* criminal. In his groundbreaking 1984 book *Hackers: Heroes of the Computer Revolution,* author Steven Levy chronicled the rise of hacker culture and described the core principles of hacking, which he labeled the "hacker ethic":

- Access to computers—and anything which might teach you something about the way the world works—should be unlimited and total.
- All information should be free.
- People should mistrust authority, and promote decentralization.
- Hackers should be judged by their hacking, not criteria such as degrees, age, race, or position.
- Art and beauty can be created on a computer.
- Computers can change a person's life for the better.

The goal of hacking in those early days was to test the limits of a system, figure out how it worked from the inside out, and—if possible—make it better. At its best, hacking was about the pursuit of knowledge for its own sake and the redistribution of intellectual wealth. In the techie's version of the Hippocratic oath, a good hack first did no harm. All in all, not bad qualities to inculcate in our children.

Unfortunately, this electronic state of grace didn't last long. In fairly short order the hacker community began to subdivide into those who used their technical skills in the classic pursuit of knowledge and those who saw the tremendous economic potential in being able to penetrate, or "crack," insecure systems. In today's nomenclature, those in the second category are the Black Hats. The good guys are the White Hats, and in between are the Gray Hats—hackers who, for instance, might break into a system to alert its administrators of weaknesses, then offer to fix those weaknesses for a fee. At the top of the hacker food chain are the so-called Elites, or "l33ts" in so-called "h4xor sp33k" (for a translation, visit UrbanDictionary.com, which will provide you with hours of eye-opening insights into contemporary slang). At the bottom are so-called "noobs" and "scrip kiddies," individuals (typically teens) who have little practical knowledge of how computer systems and networks actually operate. Nonetheless, they are able to use automated software tools, called "scripts," that give them the ability to do some low-level hacking.

That last point is particularly relevant to parents. Many of you will no doubt be tempted to dismiss this section as inapplicable to your child, because you're certain he or she doesn't have the technical skills needed to hack into a computer system. But be aware: all a child has to do is enter the phrase "hacking tools" into Google, and the search engine will instantaneously offer up thirty-seven million hits. Simply put, it's child's play to begin cracking into computer systems.

Neither state nor federal legislators have shown much sympathy for the original ideals of hacking, with its carefree explorations of computer systems and their capabilities. Virtually every state has adopted a statute that makes it a crime to commit unauthorized computer access (sometimes called computer trespass), computer fraud, theft of electronic information, or computer damage, no matter how high-minded the perpetrator's ideals might be. Federal law also prohibits computer fraud, along with other types of unauthorized intrusions.

2. The Scope of the Problem

As with other types of computer crimes, there are no clear statistics regarding the number of unauthorized electronic intrusions that occur in this country. In large part, that's because the definition of what constitutes unauthorized access, or "hacking," is so broad that it covers everything from gaining root access to a bank network to changing a sibling's Facebook photo. Moreover, many victims of hacking, from individuals to the largest institutions, don't report incidents, either because they're embarrassed or because they don't want to reveal vulnerabilities that could be exploited by others. Given how often the term "teen hacker" shows up in news headlines, however, it's a safe bet that a statistically significant percentage of all hacking is committed by individuals under the age of eighteen.

There have been a few attempts to measure the incidence of teen hacking through surveys, but as with any study that asks about criminal activity, there should be some skepticism as to whether people answered the questions honestly. In 2007, an anonymous survey of 4,800 high school students in San Diego found that one in five had accessed a computer or website without permission, and 16 percent of those took material without permission. Roughly 13 percent said they'd changed a computer system, file program, or website without permission. Of dubious comfort was the finding that only one in ten intruders did so for money or to make trouble; the bulk of the students said they broke in to learn more about how the computers worked or because the intrusion was "exciting and challenging."

In April 2010, a similar survey was conducted of one thousand high school students in New York. The questions were slightly different, so it's hard to compare results, but in that survey, one in six students admitted to hacking. The most common reasons were "for fun" (54 percent) and "out of curiosity" (30 percent). Less salutary reasons were cited as well: "to be disruptive" (14 percent), "to earn money" (7 percent), and "to have a hacker career" (6 percent).

One thing the New York survey does make clear: hacking is an activity that can be enjoyed by very young kids. Of those who admitted to hacking, just under 40 percent said they began doing so by the age of thirteen.

There's an old saying, attributed to the well-known confidence man Canada Bill Jones: "It's immoral to let a sucker keep his money." Some of the hacks teens perpetrate seem to honor that spirit—breaking into computer systems so badly protected that they've essentially put out an electronic welcome mat. For instance, in 2008, a fifteen-year-old high school student in Little Chute, Wisconsin, gained a certain notoriety for breaking into his school's computer network using a copy of the book *Internet for Dummies* as a guide. The consequences, admittedly, were serious, since it cost the school district several thousand dollars to shore up its security, but to a certain degree one could argue that the damages were self-inflicted.

Far less amusing, of course, are the hacks that cause severe system damage, significant financial losses, or threaten bodily injury. A particularly disturbing example of teen hacking occurred in Massachusetts in 2004 and 2005, when an unidentified juvenile pled guilty to "three counts of making bomb threats against a person or property, three counts of causing damage to a protected computer system, two counts of wire fraud, one count of aggravated identity theft, and one count of obtaining information from a protected computer in furtherance of a criminal act." The charges were based on a variety of incidents, including phoned-in bomb threats to schools and businesses; hacks of multiple organizations; and the downloading and online posting of confidential information leading to identity theft and computer fraud. According to a press release issued by the U.S. Attorney, the juvenile caused more than $1 million in damages to the people he victimized.

3. Consequences

The wording of each state's computer-intrusion law is slightly different, so if your child is facing charges under these statutes, you'll need to consult with an attorney in your state. In most states, any unauthorized access is sufficient to support a misdemeanor conviction; felony convictions are reserved for intrusions where the hacker had the intent to commit some additional crime, such as the theft of financial information. In a smaller number of states, mere access a computer without authorization is not sufficient for a conviction of any kind; there must also be an intent to defraud someone or steal information.

Another issue states consider is the amount of damage done by the unauthorized intrusion. Typically, there is a threshold amount that tips the crime from a misdemeanor to a felony. So, for instance, in Alabama, damages from unauthorized intrusion below $2,500 result in a misdemeanor charge, while anything above that threshold is treated as a felony. Ditto for Arkansas. In Delaware, felony prosecution kicks in at $500; in Iowa, it's a Class-D felony for damages between $1,000 and $10,000, and a Class-C felony for damages beyond $10,000.

4. Investigation and Prevention

The common thread that runs through these instances of teen identity theft and hacking is the sheer amount of available time the teens were able to put toward developing their skills and committing their crimes. Shiva Sharma was an active identity thief for years, and it took the unnamed Massachusetts juvenile well over a year to commit all his computer-related crimes. There's probably no better case for taking a long, hard look at how much time your child spends online; unlike some of the cybertraps in this book—such as cyberbullying or sexting—which can occur in seconds, effective identity theft and hacking both take serious time and effort.

Obviously, an important place to start with your preventative measures is a healthy dose of cyberethics (or just plain ethics). Hacking may look like a victimless crime, but it rarely is. Even if your child does not intend to do any harm, there is a very real possibility that he or she will do something inadvertently that will delete data, crash a system or network, or in extreme cases, cause physical damage to the computer system itself or to something (or someone) that the computer system is supposed to be monitoring. Consider, for instance, what might happen if your child hacks into a hospital network and accidentally shuts down a life-support system. Periodic conversations with your child about your expectations, your values, and the legal consequences of a hacking prosecution and/or conviction can be persuasive.

As with identity theft, episodic inspections of your child's computer are important, but if your child has a high level of computer skills, it will be all too easy for him or her to hide traces of hacking activity. Surveillance software is the next logical option, but again, depending on your child's skill set, that may not be effective.

If you have serious concerns that your child may be engaged in potentially harmful hacking activity, then the best solution is seeking outside assistance and, depending on what you learn, placing strict limitations on electronic activity. The type of outside help you hire will depend on your child's skill level. If he or she is not that experienced, a local computer technician can probably give you a good idea of what your child is doing. But if your child is highly skilled, you may want to have a computer forensics expert conduct a thorough examination of all the activity on the computer or mobile device. It's not a cheap option, but undoubtedly less expensive than a criminal prosecution and/or conviction.

Federal Computer Fraud and Abuse, and Cyberterrorism
1. The Crimes

Recognizing a surge in computer-related crimes, Congress in 1986 passed the Computer Fraud and Abuse Act (CFAA) (18 U.S.C. § 1030), which makes it a federal felony to crack into a "protected computer" for various illegal purposes. A protected computer is defined by the law as one in use by the U.S. government or a financial institution, or one used in interstate commerce or communication (which is actually a pretty large percentage of all business computers).

Among the acts which can trigger prosecution under the CFAA:

- accessing a protected computer without authorization to obtain other certain types of information, including financial records, consumer data, or information "from any department or agency of the United States";

- accessing a protected computer without authorization to obtain national security secrets;

- accessing a protected computer without authorization and affecting the government's use of the computer (i.e., crashing the computer, damaging a website, or running a distributed denial-of-service attack);

- accessing a protected computer with the intent to defraud;

- transmitting software or code to a protected computer that causes damages to a person or persons in excess of $5,000, physical injury, a threat to public safety, damage to a government computer system, or the alteration or potential alteration of medical information; and

- knowingly, and with intent to defraud, trafficking in passwords (for instance, on a cracking website) that can be used to access a protected computer without authorization.

In simpler terms, the purpose of the federal computer fraud law is to prohibit unauthorized access of sensitive computers, either for fun or for commercial gain, as well as the theft of confidential information from such computers.

The Department of Homeland Security defines cyberterrorism as "a criminal act perpetrated through computers resulting in violence, death and/or destruction, and creating terror for the purpose of coercing a government to change its policies." While there is currently no federal law on the books specifically addressing cyberterrorism, any overt acts along those lines would likely be prosecuted under the Computer Abuse statute, or other relevant statutes, depending on the underlying nature of the crime. And given the growing concern over the possibility of cyberattacks against the United States—from both within and without—it's likely that Congress will take on the issue of cyberterrorism in the near future.

2. The Scope of the Problem

As the Massachusetts hacking case illustrates, the federal government will not hesitate to prosecute teenagers for computer-related crimes. Precisely how often such crimes occur is unclear, but the number of examples and case studies is seemingly endless. The raft of prosecutions offer two chief lessons: first, that the federal government takes computer hacking and computer abuse very seriously; and second, that there are long-standing and disturbing vulnerabilities in our government's computer systems. Here are just two illustrative examples:

In 2000, a sixteen-year-old was sentenced to six months in federal prison after pleading guilty to breaking into a military computer network at the Defense Threat Reduction Agency, a government facility that guards against "nuclear, chemical, biological and conventional [but apparently not computer] weapon attacks." He also hacked into twelve NASA computers at the Marshall Space Flight Center and stole $1.7 million in proprietary software. The theft resulted in a three-week shutdown of the NASA computer system. According to then-Attorney General Janet Reno, the boy was the first juvenile sent to federal prison for hacking.

He's hardly the only teen to crack his way into sensitive governmental computers. In a series of events straight out of the movie *War Games,* sixteen-year-old Joseph McElroy, a British student, in 2004 wrote a program called "Deathserv" that he used to bypass online security at the U.S. Fermi National Accelerator Laboratory in Illinois. McElroy's motivation was almost farcical: he wanted to make use of the Fermilab's advanced computer network to provide him with super-high-speed downloads of pirated music and movie files. Whether through design or accident, other hackers obtained copies of "Deathserv," and so many used the software that the Fermi-lab's network slowed to a crawl. When U.S. Department of Energy computer experts detected the unauthorized access, they initially interpreted it as a terrorist attack and shut down the lab's computer for three days; the DOE also issued a full-scale cyberterrorist alert. McElroy was sentenced to two hundred hours of community service by a British judge.

There are myriad ways in which teens can get into trouble with the federal government through computer abuse, but there is one that is particularly rampant right now, called a "denial-of-service" or "distributed-denial-of-service" attack. The goal of the attack is to crash a target computer system by swamping it with requests for pages and other information, to the point that the website or web server slows to a crawl or simply stops working. As with other types of hacks, the basic information on how to launch a DoS or DDoS attack—and the necessary software—are all readily available online, making it a crime that even the most inexperienced hacker can commit.

An interesting twist on this phenomenon has been occurring lately: the rise of "hacktivism," where idealistic individuals (many of them children) use hacking techniques and attack software to express their disapproval of government or corporate policies. In Holland, for instance, a sixteen-year-old who assisted in denial-of-service attacks on several financial services companies, including MasterCard and PayPal, was arrested on December 10, 2010. According to the Dutch national prosecutor's office, the youth was reportedly upset that the companies had stopped processing donations for the WikiLeaks website following its release of U.S. diplomatic information.

The young man is reportedly a member of an online group called Anonymous, a highly secretive organization of renegade computer users. The group styles itself after V, the rebellious protagonist in the graphic novel (and subsequent film) *V for Vendetta,* and reportedly maintains numerous teen members in countries around the world. The attack on the financial services companies was launched using a program distributed by Anonymous called "Low Impact Ion Cannon," essentially a plug-and-play DoS generator that can be turned on by remote command when Anonymous determines there's a worthy target.

3. Consequences

Regardless of how well-intentioned your child's hacking activities may be, the consequences of a prosecution and conviction for computer abuse are potentially severe. The penalties vary depending on the seriousness of the crime and the jurisdiction in question, but as a rule, your child will be faced with the possibility of both a substantial fine and/or imprisonment. Under 18 U.S.C. § 1030, for instance, jail sentences range from one to ten years depending on the specifics of the crime, and there can be daunting fines. In the case of the Dutch teen who was arrested for using software to launch a distributed denial of service attack on credit card sites that stopped processing donations to Wikileaks, he could face a maximum of six years in prison. However, the fact that he quickly confessed to his role in the Internet attacks and his cooperation with authorities will probably significantly reduce his sentence.

It's important to note that the most severe penalties—prison time and substantial fines—are logically reserved for the most serious computer assaults. If your child's burgeoning hacking career is more for fun than profit, and causes little or no damage, non-incarceration penalties are more likely: community service, probation, reparation, and so forth. But what makes hacking so dangerous is that there are so many tools readily available to teens that can do unprecedented levels of damage to computer systems, networks, and even chunks of the Internet itself.

A good example of this is the so-called "Morris worm," a short program written by then-Cornell student Robert Tappan Morris in 1998. The purpose of the program was to gauge how big the Internet had become, but Morris made a small error that enabled multiple copies of his program to be loaded onto infected systems. The worm spread rapidly and caused a major slowdown of the Internet. According to the U.S. Government Accountability Office, the damage caused by the teenager's program was somewhere between $10 million and $100 million. Morris became the first person convicted under the federal Computer Fraud and Abuse Act; he was sentenced to three years probation, four hundred hours of community service, and a fine of $10,000 (though he eventually overcame his infamy and got a job at MIT as a professor in the Electrical Engineering and Computer Science department).

Today, much more destructive software can be downloaded and run by teens who don't have a fraction of Morris's computer programming skills.

4. Investigation and Prevention

If you think there's a possibility your child is spending time on hacker websites, or is using hacking software tools—or might even be a member of Anonymous—some drastic levels of investigation may be necessary. Just as your child doesn't have to have exceptional computer skills to cause potentially significant damage though hacking (or to be prosecuted for doing so), he or she probably has access to all the information necessary to sniff out and remove surveillance software and otherwise block any discovery of risky and illegal online activities.

If you install surveillance software and it suddenly stops working, or if it serves up an endless stream of banal information that doesn't match your gut instinct about your child's anti-establishment attitudes and computer interests, you may need to think about hiring a computer technician or even a forensics professional to periodically do a much more detailed analysis of the activity taking place on the computer. That still may not reveal everything—it's possible for a dedicated and fairly skilled computer user to wipe out virtually all traces of what they've been doing, or, as a last resort, irrevocably dispose of a hard drive or other computer storage device.

Ideally, though, things haven't yet reached that level, and you're reading this book long before your child is thinking about activities like hacking MasterCard. If begun early enough, the most effective preventative measures are lessons in cyberethics and balance. The common threads running through the stories of young people caught up in the most serious criminal cybertraps are the amount of time the children spent on their computers and the fact that, in most cases, the computer was their only significant activity.

Cyberethics is a subset of broader ethical discussion about acceptable behavior. Presumably your child has been taught, for instance, that it is illegal and wrong to smash the windows of a bank or set it on fire. It's equally wrong to try to shut down the bank's web server by launching a denial-of-service attack, but, as with most of the cybertraps discussed in this book, that's a more difficult ethical lesson for parents to teach. In part, that's because it's sometimes hard for parents to familiarize themselves enough with the technology to discuss it effectively, but also because the damage done by these activities can seem remote and not immediately visible—a child involved in hacking, unlike a child wielding a brick, doesn't have to be directly confronted with the consequences of his or her actions. Moreover, software tools like Anonymous's Low Impact Ion Cannon magnify the power of the individual by dramatically increasing the damage one person can do.

Despite the difficulty, we all need to share these lessons with our children. We also need to teach them balance, that it is not healthy to spend all of one's leisure time at the computer.

Balance, particularly between computer and physical activity, is something that can benefit us all, but it's particularly important for children. A life spent predominantly online is a life increasingly untethered from the real world. It is a life governed more by the curious and shifting mores of the Internet and less by the ethical and moral strictures of communal human interaction. Helping children achieve and maintain a proper balance in their day-to-day activities is a challenging task, given that adult struggles to do the same, but it can make a huge difference in your child's life.

Chapter Ten
Obscenity and Child Pornography

Obscenity

In the early 1870s, a young sales clerk named Anthony Comstock was hired by the YMCA to help stamp out obscenity on the sordid streets of New York City. As part of that mission, in the winter of 1873 YMCA leaders sent Comstock to Washington, D.C., to lobby Congress for a broad federal anti-obscenity law. Comstock's pleas initially fell on deaf ears, as the outgoing Congress was too embroiled in its own scandals to pay him much mind. In the waning hours of the legislative session, however—just before adjourning—lawmakers managed to pass his bill, which had been drafted with the help of U.S. Supreme Court Justice William Strong. The so-called Comstock Act, among its other provisions, made it illegal to send obscene materials through the U.S. mail. Before he left Washington, Comstock was given an appointment as a special agent for the U.S. Postal Service, which gave him the authority to search for and seize obscenity in the mail. It was a job Comstock undertook with enormous relish, and one he stuck with for the remaining forty-two years of his life.

1. The Crime

In the years since Comstock's initial lobbying effort, Congress has periodically extended the law that bears his name to keep pace with changing technology. In the early part of the twentieth century, for instance, Congress passed the Radio Act of 1927 and the Communications Act of 1934, both of which used language from the Comstock Act to set restrictions on what could be broadcast over the airways. In 1968, the Comstock Act was extended to cover obscene phone calls, including those that used "lewd, lascivious, filthy, or indecent language." And in 1996, in a panic over the rise of the online adult industry and its perceived impact on children, Congress passed the Communications Decency Act, which applied Comstock's obscenity language to the Internet. That law also attempted to prohibit the transmission of "indecent"—not just obscene—materials, but that provision was struck down by a unanimous 1997 Supreme Court decision.

All these laws are in tension, of course, with the provisions of the First Amendment, which—with limited exceptions—broadly protects speech, even speech that some might find offensive or distasteful. Federal and state attorneys who want to prosecute obscenity have the burden of first proving to a jury that it is, in fact, obscene. The final word on the legal standard for obscenity has fallen to the Supreme Court, which in a 1973 case—Miller v. California—set a three-part test for determining whether a particular work (a book, poem, song, movie, statue, etc.) could be prohibited:

1. whether "the average person, applying contemporary community standards" would find that the work appeals to prurient interests; and

2. whether the work depicts or describes, in a patently offensive way, sexual conduct specifically defined by applicable state laws; and

3. whether the work, when taken as a whole, lacks serious literary, artistic, political, or scientific value.

If a jury answers all three questions in the affirmative, the speech in question is deemed obscene and is not protected by the First Amendment.

When it comes to online obscenity, it's a crime—punishable by up to five years in prison—to receive from any interactive computer service (i.e., the Internet) "(a) any obscene, lewd, lascivious, or filthy book, pamphlet, picture, motion-picture film, paper, letter, writing, print, or other matter of indecent character; or (b) any obscene, lewd, lascivious, or filthy phonograph recording, electrical transcription, or other article or thing capable of producing sound" (18 U.S.C. § 1462). The same penalty can apply to obscene work produced "with the intent to transport, distribute, or transmit in interstate or foreign commerce," including through the use of an interactive computer service (18 U.S.C. § 1465).

In addition to the federal statutes, each state has its own obscenity laws. However, the standard for obscenity under state law cannot be stricter than the one set forth in the Miller case.

2. The Scope of the Problem

The original purpose of both federal and state obscenity laws was to punish individuals who produced, published, and distributed obscene materials. By and large, that has limited obscenity prosecutions to adults, who are more likely than children or teens to have the means and opportunity to publish adult movies and videos—or, more recently, adult websites. In the 1960s there were several prosecutions of high school students for producing allegedly obscene underground newspapers, but obscenity cases involving teens have been rare. Over the last forty years, there have been virtually no obscenity prosecutions based solely on written materials.

In a handful of instances, teens and preteens have been charged under state obscenity laws for other types of acts. For instance, in 1998, a Texas high school student was arrested and charged with obscenity for wearing a particularly profane Marilyn Manson t-shirt in public (the case was later thrown out). In 2007, five Louisiana fifth-graders were charged under the state obscenity law for having sex in front of classmates while their teacher was out of the room. And in August 2010, North Carolina officials charged nineteen-year-old James Lackey with obscenity for sending a photo of his genitals to a thirty-nine-year-old woman; Lackey told authorities he had intended the image as a joke for a friend, and had punched in the wrong phone number. He was released on an unsecured bond, but there were no further news reports on whether his excuse was accepted.

Thanks to the potentially toxic mix of the Internet and camera phones, there has been a recent surge in state and federal obscenity charges against teens and even preteens who have taken and distributed nude photos of themselves and others, on the grounds that such images constitute child pornography. It's possible your child could face more traditional types of obscenity charges for his or her online activity, but in reality the likelihood is relatively low. When I was researching my first book, *Obscene Profits,* there were stories of high school students signing up as affiliates of adult websites to distribute explicit content, but the industry worked hard to weed them out and none were prosecuted. In fact, in the sixteen years or so since the Internet became a fixture of American life, there have been very few online obscenity prosecutions (much to the dismay of social and religious conservatives).

Nonetheless, if your child was running an adult website from his or her bedroom—and given the amount of free content online, it wouldn't be that hard to set one up—a prosecutor could conceivably pursue federal obscenity charges. Similarly, your child could be prosecuted for obscenity if he or she was producing and distributing homemade adult videos, or even for child pornography if the videos included minors. Both activities, virtually unthinkable a decade ago, are all too feasible in an era of broadband Internet access, web cams, and high-definition Flip video cameras (or the iPhone 4, which includes video capability).

3. Consequences

Of the cybertraps discussed in this book, the possibility of a traditional obscenity prosecution is probably the least likely to snag your child. There are a variety of reasons why the odds are so low. First, the application of obscenity laws to the Internet is mostly uncharted legal territory, and the bulk of state and federal law enforcement resources are devoted (appropriately) to the pursuit and prosecution of child pornographers, where there are far fewer legal ambiguities. Second, to the extent that officials are prosecuting obscenity, they are focusing their attentions on producers and distributors, not mere possessors. Third, even if your child did decide to launch an adult website, the chances of anyone noticing would be remote, given the sheer amount of sexually explicit material online. Finally, even a moderately savvy teen would quickly realize that running an adult website is a waste of time; with all the free content available at the click of a mouse, only a small number of sites actually make money.

Still, it's a good idea to know if your child is downloading potentially obscene adult materials (not only for legal liability, of course, but for his or her emotional and moral well-being). If your child is using file-sharing software, there is a substantial risk of accidentally downloading child pornography, which *is* a significant cybertrap. Moreover, that same software would immediately make your child a distributor of whatever content he or she downloaded, raising the odds of law-enforcement attention.

4. Investigation and Prevention

You can monitor your child's downloads by installing surveillance software, or by visually inspecting the computer. If you decide to search your child's computer yourself, your search should focus on visual images—typically stored with .jpg or .gif file extensions—and video files, generally stored with .mpg, .avi, or .mov file extensions. You can search your child's computer for all files of a specific type by opening Windows Explorer (or Macintosh HD if your child uses a Mac) and entering *.jpg, *.gif, or similar extension in the search box.

You should also check to see if one or more file-sharing software packages have been installed. Visit the book's website (www.cybertrapsfortheyoung.com) for details on how to do that search. But keep in mind: if your child is a sophisticated computer user, he or she may take steps to hide image or video files from easy discovery, in which case you may need to hire professional help. Also, review the options discussed in appendix A for regularly monitoring your child's electronic activity. The software packages on the market can help you identify the programs your child is using and exactly what he or she is doing with them. The key lesson here is that as the potential cybertraps grow steadily more serious, the levels of supervision and monitoring need to increase accordingly.

Child Pornography

1. The Crime

Child pornography is a subset of the broader category of obscenity. Unlike other types of speech (even sexually explicit speech that might be considered obscene), child pornography is not entitled to a presumption of protection under the First Amendment. Congress has adopted a variety of laws that explicitly define child pornography and make it a felony to possess, receive, distribute, or produce such material in a way that affects interstate or foreign commerce (18 U.S.C. §§ 2251,

2252, 2252A). To pass that legal bar, an image or set of images must have crossed state lines, or been produced using products or materials that themselves crossed state lines. If, for instance, a pornographic image of a child is downloaded from a website, it almost certainly crossed state lines. Even if the image traveled entirely within the boundaries of a single state, it was undoubtedly, at some point, stored on a hard drive or CD that was itself shipped across state lines, which would be sufficient to give the federal government jurisdiction to prosecute. Congress has defined child pornography as the following:

> Any visual depiction, including any photograph, film, video, picture, or computer or computer-generated image or picture, whether made or produced by electronic, mechanical, or other means, of sexually explicit conduct, where:
>> (A) The production of such visual depiction involves the use of a minor engaging in sexually explicit conduct;
>> (B) Such visual depiction is a digital image, computer image, or computer-generated image that is, or is indistinguishable from, that of a minor engaging in sexually explicit conduct; or
>> (C) Such visual depiction has been created, adapted, or modified to appear that an identifiable minor is engaging in sexually explicit conduct (18 U.S.C. § 2256).

The goal of this statute is not only to make it illegal to distribute sexually explicit images or videos of individuals under the age of eighteen, but also to prohibit individuals from using computer software to create human-like and sexually explicit animations of minors, or to modify sexually explicit images to give the appearance that a minor is engaged in sexually explicit activity (for instance, by Photoshopping the head of a minor on the body of a adult engaged in sexual activity).

2. Consequences

Penalties under federal child pornography statutes are severe. Conviction for the production of child pornography carries a fifteen-year mandatory minimum sentence; sentences for possession, receipt, and distribution of child pornography range from a mandatory minimum of five years in federal prison to a maximum of twenty years.

Every state also has its own laws prohibiting the production, possession, and distribution of child pornography. The specific parameters of the crime, the definition of what constitutes child pornography, and the penalties for the various acts vary from state to state, so it's important to consult with an attorney licensed to practice in your jurisdiction if your child is facing prosecution.

One of the initial questions to ask is whether your child will be tried as a juvenile or as an adult, a decision that can play a critical role in the severity of the potential sentence. Unfortunately for children, there has been a strong trend over the last fifteen years toward prosecuting minors as adults, particularly when the crime is severe or a repeat offense, or when the child is near the age of majority.

For instance, in the summer of 2010, seventeen-year-old Bryan Nguyen was arrested in Moss Point, Mississippi, for possession of child pornography. He was charged as an adult and faced up to forty years in prison if convicted. In the summer of 2009, a sixteen-year-old Tampa, Florida teen named Patrick Melton was arrested for downloading pornographic images of children and was initially charged as a minor, but after reviewing the case the state's attorney decided to instead try Melton as an adult. And in 2006, Ryan Zylstra, a seventeen-year-old Michigan teen, faced up to twenty years in prison after being prosecuted as an adult for posting a photo to his blog of two friends having sex at a New Year's party. Since the girl in the photo was only sixteen, the image qualified as child pornography. Zylstra later pled guilty to improper surveillance and was sentenced to one hundred and twenty hours of community service and two years probation.

3. Accidentally Trapped

In this particular instance, the focus is not so much on how many kids are accessing child pornography (since there's no way to quantify that), but on how easily kids can be caught in this particular cybertrap. Far too many children (and adults) simply stumble into this trap through the careless use of peer-to-peer software.

The idea behind peer-to-peer software is simple: it allows computer users to search for and download files directly from other computer users employing the same software. Until recently, the most popular P2P program was LimeWire (with nearly fifty million users), but that company was ordered to shut down by a federal court due to its role in facilitating massive amounts of copyright infringement. The shutdown of LimeWire is not likely to stop the practice of illegal file sharing, however; P2P users continue to have a wide range of software options (for a list of popular file-sharing programs, visit www. cybertrapsfortheyoung.com).

The main draw of P2P software is that it enables people to find free (i.e., stolen) copies of songs, movies, and software, a characteristic that particularly appeals to kids who frequently lack cash, or who know their parents won't buy them the ear-splitting or indecent songs they crave. The second most popular reason for using P2P is to download free pornography, which is where things can get dangerous.

All P2P software is built around a search engine that indexes files on the P2P network. When someone searches for a file with a particular name, the software looks in the index and, if it finds one or more matches, presents the searcher with a list of results. The results show the names of all the matching files and the users who are making that content available. Once a user clicks on a file name, it begins downloading to his or her computer.

If your child clicks on a file name that clearly indicates it might be child pornography, it will be that much harder to raise a defense that the download was accidental. But if your child does a bulk download of all the search results, he or she could wind up downloading child pornography without realizing it. That scenario does allow for the possibility of arguing that the receipt and possession was accidental, but juries and judges are often skeptical. Another problem is that it's extremely common for P2P files to be mislabeled, so your child might think he's downloading a popular song or movie and only after the fact discovers that he's downloaded child pornography.

By default, P2P software is set up to store files in a shared directory on the user's computer, where they are indexed and listed as available for other users of the software. If your child installs P2P software without changing the default settings, everything he or she downloads from the P2P network can be immediately retrieved by others. If any of those shared files are pornographic images or videos of children, he or she could be charged under child pornography distribution statutes, which carry more severe penalties than mere possession.

Law enforcement officers frequently use P2P software themselves to identify potential child pornography defendants, searching the networks for illegal files and recording the user IDs and IP addresses of individuals offering to share them. The odds are high that if your child downloads a file containing child pornography, the Internet Protocol address for the computer—or for your home wireless router—will be identified and logged in a law enforcement database.

4. Intentionally Trapped
It's important to acknowledge that not every child who gets snared in the child pornography cybertrap simply stumbles

into it. For a small minority of kids, it's a trap that catches them after they deliberately search for sexual images of other minors. In other cases, a child might graduate to child pornography after starting with so-called adult content. It is also a cybertrap that overlaps significantly with sexting, discussed in detail in chapter 11.

P2P software is by far the most dangerous element in these scenarios. In addition to all the other stolen content circulating on P2P networks, there is a huge amount of child pornography that is easily located with search terms like "lolita," "preteen," "pthc" (preteen hard core), or "hussyfan." If you've installed monitoring software on your child's computer, you can set it up to alert you if your child types any of these words. Unless you know your child is an earnest Nabakov fan, you'll want to do more investigation immediately if any software alarms go off. If you haven't installed monitoring software, there are also relatively inexpensive tools that will enable you to search your child's computer for specific keywords.

If you've been paying even a moderate amount of attention to the growth of the Internet over the last fifteen years, you know there is a staggering amount of sexually explicit material online. A significant portion of that content features young men and women who are barely eighteen, and such sites are aggressively marketed. For instance, if you type the phrase "teen videos" into Google, the first page and a half of results are for adult websites promising "young teen virgins," "virgin sex videos," and so on. Federal law requires websites to maintain records proving their models are over eighteen, but that statute doesn't apply to foreign sites. And in truth, there are far too many explicit websites for the law to be effectively enforced.

While you might not like your child browsing through or downloading content from the typical "barely legal" website, the truth is that doing so is not particularly likely to trigger law enforcement interest. Unlike P2P networks, which can be monitored efficiently in real time, the Web is simply too vast for law enforcement to monitor. Moreover, because the "barely legal" websites have fixed addresses and are easily found, they're less likely to traffic in the type of content that most often triggers prosecution: videos of very young children being sexually abused.

Access to that kind of content is much harder to find, and requires conscious effort. What typically happens is that an individual or group will set up a hidden website which can be used to store and trade child pornography. Access to the site will be carefully advertised in chat rooms and obscure message groups catering to individuals who are looking for preteen content, and an individual will only be admitted after he has established his bona fides by contributing new material. Law enforcement agents will often spend months working undercover to gain access to these sites, and they routinely patrol Internet chat rooms looking for indications that new hidden sites are being advertised. When the sites are infiltrated and taken down, investigators usually capture a treasure trove of user names, IP addresses, and credit card numbers to assist in making arrests. It's not uncommon for these raids to result in the wholesale arrest of hundreds of people around the country at one time. There are very few minors who have been arrested for such activities, but it certainly could happen.

The last scenario in which a child could be prosecuted for child pornography involves the production and distribution of sexually explicit images and videos involving other minors, or themselves. Some of these charges are overkill for what is really just a careless or naive misuse of technology, but there have been incidents that suggest kids can sometimes have commercial motives. For instance, in 2009, Vermont officials brought child pornography charges against a South Burlington High School student who was filming sexually explicit movies using his mobile phone. Although there was no evidence that any of the videos had been sold, they had a staged quality to them that suggested much more intent than the typical opportunistic video of sexual activity at a party.

5. Investigation and Prevention

From an investigative point of view, there's one relatively simple step you can take that will significantly cut down the chances of your child falling into the child pornography trap: check to see if your child has installed any P2P software (including software that downloads a kind of file called bit torrents) on the computers in your home.

If the answer is yes, my firm recommendation is to promptly uninstall it yourself or have your child do it under your immediate supervision (which will afford you the opportunity to discuss the potential risks with your child as he or she does so). There are legal uses for P2P software, but they are relatively few and far between, particularly compared to the large number of illegal uses. And those legal uses are dwarfed by the risk of criminal prosecution. Instructions for searching your child's computer for these programs can be found on the book's website at www.cybertrapsfortheyoung.com.

The next step is to seriously consider installing software to monitor what your child is doing online. While P2P software is by far the most common way that children are exposed to child pornography, it's hardly the only way, and surveillance software can help you keep track of all the websites and chat rooms your child visits. They can alert you if your child uses certain keywords while online, and you can receive a report listing the programs your child uses on the computer. Depending on your schedule and your own level of familiarity with computers, this may be the most practical tool for protecting your child.

As with the other cybertraps discussed in this book, prevention is the best approach, and your most powerful tools are communication and education. As you educate your child about what constitutes appropriate behavior concerning his or her body, it is appropriate to add in some admonitions about the use of electronics, even at an early age. After all, if your toddler or elementary-aged child has a digital camera (and the odds are very good he or she at least has access to one), it's important to make clear the appropriate boundaries regarding its use. A child may not be old enough to upload photos to the Web or send a text message, but kids invariably gain that knowledge—and sooner than we think, particularly if there are older children in the house.

It is equally important to have conversations with your child about sexual issues, and the presence of sexual material on the Internet. Explain that while curiosity about sexual matters is perfectly normal, satisfying that curiosity online is dangerous. Tell your child that there are numerous examples of teens and young adults who have been prosecuted and sent to prison for downloading just a handful of sexually explicit images of minors. Encourage your child to tell you promptly, without fear of punishment, if a classmate or someone else forwards a sexually suggestive or explicit image or video. This last point is analogous to the agreement many parents have with their children that there will be no repercussions if they call for a ride in a situation where all the potential drivers have been drinking.

Parents often wonder how early these types of conversations should begin. Given how easy it is to install and use file-sharing software, and given the severity of the consequences, it's appropriate to begin talking about these issues as your child is entering middle school—or perhaps even earlier, depending on how much time they spend on the computer. These aren't comfortable conversations to have, but they're necessary ones. If your child is going to be cruising down the information superhighway, he or she needs to know just how serious the speed traps can be.

Chapter Eleven
Sexting and Sextortion

Sexting

I first became aware of the "sexting" phenomenon in January 2009, when Isaac Owusu, a seventeen-year-old student in South Burlington (one town over from mine) was suspended for sexually assaulting two high school girls. Naturally, the crime itself attracted attention, but what really raised eyebrows was the subsequent discovery that a number of girls in South Burlington High School had used their mobile phones to send nude photos of themselves to Owusu. Interviews with other students made it clear this was something kids had known about for some time. "If you were to check every guy's phone, you'd probably find [nude photos] on ninety-seven percent" of them, freshman Rachelle Bedard told the *Burlington Free Press*.

That may be a slightly cynical view, but there's no question that at least in northern Vermont, the South Burlington case alerted parents to a rapidly spreading national phenomenon. There had been well-publicized incidents in the prior year of celebrities taking and sending nude texts (Vanessa Hudgens of *High School Musical* fame, as well as Miley "Hannah Montana" Cyrus), but like many parts of the country Vermont tends to have an "it can't happen here" mentality. The Owusu case quickly dispelled that illusion.

1. The Crime

The South Burlington case had one other immediate effect: it raised awareness that many children around the country were being prosecuted under child pornography statutes for distributing nude or sexually explicit photos of minors (in many cases, themselves). A significant number of those children were prosecuted as adults and, when convicted, faced imprisonment and lengthy (if not lifelong) designations as sex offenders.

There's certainly a good argument that many sext images meet the technical definition of child pornography, and it's not surprising that some prosecutors, hoping to send a message to those involved, would charge teens under child pornography statutes. But it didn't take long for people to start asking whether it was necessary or appropriate to pin the sex-offender label on teens—particularly given all the concomitant loss of educational, employment, and living options—for engaging in what is often little more than a high-tech version of "you show me yours, I'll show you mine."

It's important to be absolutely clear that child exploitation and abuse are profoundly serious problems and ones that should receive law enforcement's best detection and prevention efforts. Likewise, a teen who betrays a partner's trust by forwarding private images to others, or by posting images to the Web, should be treated as a distributor of child pornography. I know several law enforcement officers who argue that when any of these images escape into the wild, as they invariably seem to do, they help feed the overall problem of child pornography, and thus the punishment should match. But assuming the photos are taken and shared by two teens in a non-exploitative and non-coercive relationship, saddling one or both with the label of sex offender seems too heavy-handed.

In the spring of 2009, a bill was proposed in the Vermont Senate that would have created an exception to the state's child pornography statute for the exchange of images by individuals between the ages of thirteen and eighteen. The proposal received a lot of negative attention from the national media, which interpreted the Senate language as implicitly endorsing the practice of sexting.

The Vermont House Judiciary Committee proposed amended language, later adopted by both chambers, that created a new crime called Minor Electronically Disseminating Indecent Material to Another Person (13 V.S.A. § 2802b). For a first offense, the matter is handled in juvenile court and the state cannot bring child pornography charges (which eliminates the possibility of sex-offender registration). If it's a second or subsequent offense, the choice between juvenile or criminal court is left to the discretion of the prosecutor. In addition, if an individual over the age of eighteen is found in possession of a sexting image, he or she can be imprisoned for up to six months. A prosecutor also has discretion to bring other sexual misconduct charges if appropriate.

Over the last couple years, ten states have adopted similar legislation, modifying their child pornography statutes to lessen the consequences of teen sexting (for a complete list of those states and links to pending legislation, visit www.cybertraps-fortheyoung.com). The remaining forty states (and the federal government) must still decide on a case-by-case basis whether the potentially severe penalties for a child pornography conviction are justified.

2. The Scope of the Problem

One good thing about a social problem with a high media profile is that people tend to conduct surveys about it. When it comes to sexting, survey results are a bit scattered, but they clearly indicate that the practice is, in fact, a significant phenomenon among teens and preteens.

In the fall of 2008, the National Campaign to Prevent Teen and Unplanned Pregnancy (NCPTUP) surveyed 653 teens between the ages of thirteen and nineteen. Among the significant results:

- 20 percent of teens aged thirteen to nineteen had sent or posted nude or semi-nude photos or videos of themselves.
- 39 percent of all teens had sent or posted sexually explicit messages.
- 11 percent of girls between thirteen and sixteen had sent or posted suggestive photos of themselves. 71 percent of those said they'd sent the images to a boyfriend.
- 21 percent of those girls, meanwhile, reported sending suggestive content to someone they wanted to date or "hook up" with. And 15 percent said they'd sent that content to someone they knew only online.

In September 2009, the Associated Press and MTV combined on a similar study that interviewed roughly 1,200 people between the ages of fourteen and twenty-four:

- 24 percent of teens aged fourteen to nineteen said they'd been involved in some form of naked sexting, either as a sender or receiver.
- Only 10 percent reported sharing a naked photo of themselves; girls were more likely to have done so (13 percent) than boys (9 percent).
- Sexually active children (those reporting they'd had sex in the previous week) were more likely to sext: 45 percent reported doing so.
- Almost a third (29 percent) of children sent sexts to people they knew only online.
- 24 percent sent sexts to people they wanted to date or "hook up" with.

Finally, the most recent comprehensive study on the sexting issue comes from the Pew Internet and American Life Project, which in the summer of 2009 conducted a telephone survey of eight hundred teens aged twelve to seventeen, followed by a series of focus group interviews. The findings are slightly more encouraging for parents than the earlier surveys:

- Only 4 percent of mobile phone-owning teens reported sending a sexually suggestive nude or nearly nude photo or video of themselves to someone else.
- There was no distinction between genders, and the likelihood that a teen had sexted increased with age (8 percent of seventeen-year-olds said they'd done so).
- Just under 17 percent of teens said they'd received a sexually suggestive image or video of someone they knew.
- The likelihood of receiving a sexually suggestive image or video also increased with age (4 percent of twelve-year-olds, compared with 20 percent of sixteen-year-olds and 30 percent of seventeen-year-olds).

As always, it's good to keep in mind the caveat that Pew includes with its survey results: "[S]exting is a topic with a relatively high level of social disapproval. This raises the possibility that any time any researcher asks questions about the subject that respondents will not admit to engaging in the socially subject behavior, which may result in findings that underreport the actual incidence of a behavior." It may well be that teens find it somewhat easier to be honest in a written survey, or perhaps teens are awakening to the serious legal consequences. In either case, I suspect that the Pew results are low.

3. Consequences

As we all know, the practice of "kiss and tell" has been around a lot longer than modern technology. A kid doesn't need a smartphone to embarrass someone by describing a sexual encounter or revealing intimate details to others. But our modern electronic gadgets have unquestionably exacerbated the problem, partly because they increase the likelihood that images will be part of the mix, and in part because the disclosures can be shared with so many people so quickly.

Sexting is a serious cybertrap all by itself. In the majority of states, and under federal law, if sexting involves visual images the sender and receiver can both be subject to child pornography laws. But the potential harm doesn't stop there; there's not just legal fallout involved with sexting, but emotional fallout. The remarkable ease with which electronic images can be distributed worldwide, the vast range of online communication tools, and the all-too-normal tendency of teens and preteens to harass each other can turn explicit images into powerful emotional weapons. Teens who are bullied and cyberbullied as a result of widespread redistribution of their private images can suffer severe emotional trauma, physical injury, and even death.

Recent surveys of teen sexting behavior offer insights into not only how often images are redistributed, but also the motivations for that behavior. The MTV survey, for instance, found that nearly one in five teens had forwarded sexually suggestive material with someone other than the sender and, of those, 55 percent said they had shared that material with more than one person. That's not terribly surprising; after all, it only takes a few clicks or keystrokes to select multiple recipients on a smart phone.

Teens told MTV that their top reasons for sharing the content were: the assumption that others wanted to see them (52 percent), a desire to show off or boast (35 percent), as a joke (31 percent), and good old-fashioned boredom (26 percent). Kids are not completely clueless about how easily explicit content can be shared. In the NCPTUP survey, 44 percent of respondents said they knew such content was commonly shared with someone other than the intended recipient. That's significantly higher than MTV's survey, in which just 14 percent of teens who sent a nude video or image reported that they thought it would be shared with someone else.

The real problem is that teens and preteens have a difficult time appreciating how these types of images can be misused by their peers, much in the way the kids in my own high school failed to appreciate the dangers of drinking and driving until we had to attend three classmates' funerals during our senior year. Tragically, sexting-related cyberbullying is following the same narrative arc.

In the spring of 2008, Jessica Logan, a senior at Sycamore High School in Cincinnati, traveled to Florida on spring break with two friends. During their vacation, the three girls took nude photos of themselves using their mobile phones, and Jessica sent hers to her boyfriend, Ryan Salyers. A couple months later, Jessica learned from her best friend that copies of the photo were being circulated around the school. She promptly went to the school resource officer, who told her that because she was eighteen at the time she took the photo, there was nothing he could do.

Jessica's parents later learned that their daughter's nude photo had been distributed to students in at least six other high schools in the Cincinnati area. Throughout the spring, she was taunted and physically abused by students at Sycamore High School and elsewhere: she received text messages calling her "slut, porn queen, whore," and she found wall posts and messages expressing similar epithets posted to her Facebook and MySpace pages. At the suggestion of the school resource officer, Jessica did an anonymous television interview about the dangers and consequences of taking and sending nude photos. But in July 2008, just a few weeks after a fellow classmate committed suicide, Jessica hanged herself in her bedroom.

The fact that her suicide followed so closely on the heels of her classmate's led some to suggest her case was one of "suicide contagion" rather than victimization through cyberbullying.

But Jessica's parents are convinced it was the vicious treatment she received after her nude photo circulated that pushed her past the point of no return. The Logans have become vocal advocates for clearer anti-sexting laws and better education to warn children about the dangers. In 2009, they also filed a civil suit against a number of individuals and entities, including Sycamore High School. The claims run the gamut of the cybertraps discussed in this book: stalking, harassment, dissemination of private material, intentional infliction of emotional distress, negligence, breach of expectation of privacy, violation of Jessica's civil rights, and wrongful death. The Logans sought unspecified damages in an amount exceeding $25,000.

In the spring of 2009, an eighth grader in Florida named Hope Witsell sent a topless photo of herself to her boyfriend. The photo was circulated to other kids in her school, who began taunting her about it. At the start of the following school year, some of her classmates created a "Hope Hater" page to further harass her. Soon after, in an incident difficult for any parent to imagine, Hope hanged herself from her canopy bed. Her mother has started a group called Hope's Warriors to help educate other parents, and warns that, "It happened to my daughter, it can happen to yours, too. No one is untouchable."

There are many factors that contribute to teen suicide, and it's difficult to ascribe any tragedy to a single cause. But there's little doubt that the relentless harassment these teens faced when their private photos became public contributed significantly to the emotional stress that made suicide a possibility. It can't be emphasized enough: our children live in a perpetually connected world; if things go bad, there's often no refuge and no relief.

4. Investigation and Prevention
As a parent, your primary goal should be to help your teen and preteen children understand why they shouldn't join the

roughly 20 to 30 percent of teens who take and share nude photos of themselves. Nor should they allow other teens to take explicit photos of them—no matter what the reason. Parents often ask me how young I think these conversations should begin, and my reply is straightforward: does your child have a cell phone with a camera? If the answer is yes, then the time to have that conversation is right now, regardless of your child's age. Guidance about what should remain private is particularly vital for today's children. Remember, these kids live in an environment where they're constantly led to believe that sharing everything online is a social necessity (blogs, social networking sites, photo-sharing sites, etc.) and where all too many celebrities laughingly sext their lives for public consumption.

These issues are difficult for parents not merely because they're sexual in nature, but because sexual mores change from one generation to the next. Not only are the Millennials growing up in a constantly connected world, they're also growing up in a more sexually charged environment. Thanks largely to the Internet, teens and preteens are exposed to more sexual content than any generation before them. Whether you feel it is appropriate to caution your child against taking or appearing in nude photos because it is morally wrong is a private decision. However, be prepared for some resistance on generational grounds alone. Your children might find the practical arguments more compelling: very little electronic content remains private; kids get harassed over photos like these, occasionally to the point of self-inflicted injury or death.

To be fair, many kids who sext expect that their images will remain private; they trust the person taking the photos, or the person to whom the photos are sent. It can be a hurtful and painful lesson when that trust is broken. The lesson shouldn't necessarily be that trust is a bad thing; instead, children need to learn the importance of being thoughtful about their own actions. Given how easily a nude image can be transferred from one mobile phone to another, does it make sense to send it in the first place? Ask your child how he or she would feel if a sibling's nude photo was circulating around the school. While most teens want to believe they're mature, they can easily name any number of classmates who are not, and so it's not difficult to come up with examples of people who would re-circulate a nude photo if it came into their possession. The simple point is this: the most easily controlled nude photo is the one that isn't taken in the first place.

Unfortunately, parents can talk to their kids until they're blue in the face and sexting will still occur; nude photos will continue to travel farther than originally intended. The question is how best to prevent the consequences that tend to flow from redistribution. The first and most critical step is careful observation by parents, friends, teachers, and administrators. If a student has a dramatic personality change, drops out of activities or classes, or begins harming himself or herself, a quick and strong parental response is critical before the situation spirals out of control.

As the Logans and Witsells argue their cases in court, schools must become more sensitive to the ripple effects of online activity. As a school board chair myself, I'm sympathetic to the enormous social and financial challenges schools face, and the workloads teachers and administrators carry. But when off-campus activity like sexting clearly and visibly affects a student's on-campus personality and performance, I think schools have a duty to observe and report these changes. The primary responsibility resides with parents, in terms of monitoring both their children's electronic activity and mental and emotional health, but schools should be an active partner. The single most important improvement school districts could make would be to establish a clear structure for the reporting and investigation of harassment, bullying, and sexting.

Given the fact that significant harassment and bullying occur on social networking sites, it's crucial for parents to communicate with each other about what their kids are saying and doing online. If you see or hear something that concerns you about the way kids are interacting, or if you catch wind of a cyberbullying situation, send the parents of the children involved an e-mail or Facebook message, or just pick up the phone. It is far better to be a little discomfited when talking about a possible sexting incident than to struggle for the right words of consolation at a funeral.

Sextortion

The concept of sexual blackmail is not a new one, stretching back at least as far as the reign of Nebuchadnezzar II. The Roman Catholic and Eastern Orthodox versions of the Bible include a story, in the Book of Daniel, about a beautiful young Hebrew wife named Susanna. According to the tale, one day while Susanna was bathing in her garden, two elders spied on her and told her that, unless she agreed to sleep with them, they would publicly accuse her of adultery. She refused, and was sentenced to death for being unfaithful. As she was being led away to her execution, the Lord inspired a young man named Daniel (of lion's den fame) to challenge her conviction:

"Are you such fools, you Israelites," Daniel said, "that you have condemned a daughter of Israel without any examination or ascertaining of the truth? Go back to the place of trial. For these men have borne false witness against her." He had the two men separated and examined them individually. When he asked each under what type of tree Susanna supposedly met her lover, they gave different answers. The crowd that had gathered to observe Susanna's execution killed the two lying elders instead, and praised Daniel for his wisdom.

The story of Susanna has generated a lot of great art, including outstanding paintings by Artemisia Gentileschi, Rembrandt, Rubens, Van Dyck, and a particularly creepy work by famed American artist Thomas Hart Benton, who reset the tale in the American frontier. But two and a half millennia after Susanna's trial, not much has changed about human nature. There are still individuals who try to coerce sexual favors from others. What has changed is the technology that enables people to peer into others' gardens, far more easily than the two old lechers whom Daniel exposed.

1. The Crime and its Consequences

There are a number of recent cases in which adults and teens have used sexually explicit or suggestive photos to coerce their victims into providing even more explicit photos or sexual acts, an act increasingly described as "sextortion." State and federal legislators have not yet begun to grapple with how to explicitly prohibit and punish such behavior, though that's not to say there aren't laws that can be used to prosecute sextortion. The federal government, and every state, prohibits blackmail and extortion, and this type of activity would run afoul of both sets of laws. For instance, the federal statute regarding criminal acts involving interstate communication contains relevant language:

> Whoever, with intent to extort from any person, firm, association, or corporation, any money or other thing of value, transmits in interstate or foreign commerce any communication containing any threat to injure the property or reputation of the addressee or of another or the reputation of a deceased person or any threat to accuse the addressee or any other person of a crime, shall be fined under this title or imprisoned not more than two years, or both (18 U.S.C. § 875).

In addition, if your child is blackmailing or sexually extorting someone who is under the age of eighteen, he or she could be prosecuted under child pornography and child exploitation laws. It's also likely that legislators around the country will soon begin proposing bills to specifically address this category of cybercrime.

2. The Scope of the Problem

It is far too early to have any research indicating how extensive a problem sextortion is among teens. It is fair to say, however, that it affects at most a small subset of those teens who post nude or sexually suggestive photos of themselves online. As with sexting, however, even if the actual numbers are small, the potential consequences are severe. A couple recent cases help illustrate how easily teens and preteens can become both victims and perpetrators.

One of the most infamous cases arose in the winter of 2008 in New Berlin, Wisconsin, when Anthony Stancl, an eighteen-year-old senior at Eisenhower High School, was expelled after he e-mailed a bomb threat to the school. Investigators seized Stancl's computer and, during the forensics exam, uncovered even more troubling data: thirty-nine individually labeled directories containing over three hundred nude photos of male Eisenhower High students.

The results of the subsequent investigation absolutely staggered the quiet suburban town. Investigators learned that Stancl had created fake profiles on Facebook, pretending, in each instance, to be a female student. He then sent flirtatious messages from the profiles to male students, offering to exchange nude photos. When the boys complied, Stancl would send them images of teen girls he'd culled from the Internet. He then blackmailed the boys, telling each of them that he would distribute their photos to friends, or even the entire school, unless they agreed to perform certain unspecified sexual acts with him. According to authorities, Stancl succeeded in coercing at least seven male Eisenhower students to go along with his demands.

According to the *Milwaukee Journal Sentinel,* Stancl was initially charged with "a dozen felonies, including repeated sexual assault of the same child, possession of child pornography, two counts each of second- and third-degree sexual assault, five counts of child enticement and one count of causing a bomb scare." He ultimately pled guilty to two felonies: repeated sexual assault of the same child, and third-degree sexual assault; he was sentenced to fifteen years in prison and thirteen years of supervised release.

In some extortion cases, the triggering event is an act of foolishness by the victim. In the summer of 2009, for instance, three girls got together and made the mistake of mixing alcohol and a webcam. They visited an Internet chat room and agreed to anonymous requests to flash their breasts on camera. A short time later, one of the three girls—a seventeen-year-old from Indiana—was contacted by e-mail and told that if she didn't provide more explicit images to the sender he would post screen captures of her online partying to the MySpace pages of all her friends. She initially complied with the request, but eventually reported the incident to state and federal authorities. On June 9, 2010, a federal grand jury indicted nineteen-year-old Trevor Shea on charges of child sexual exploitation. If convicted, he faces a maximum of thirty years in prison and a fine of up to $25,000.

In interviews following the Shea indictment, federal prosecutors told the Associated Press they were seeing a steady increase in the number of computer-related sextortion cases. The easy availability of nude teen sext images and webcam shots, prosecutors said, has provided perpetrators—both young and old—with ample blackmail material.

3. Investigation and Prevention

As with sexting, the primary responsibility for prevention begins at home. If your child isn't posting or sending nude images of him- or herself, the chances that he or she will become a victim of sextortion go down precipitously. But that's only half the equation: it's equally important to make sure your child is not a sextortionist. The basic requirements are education and supervision.

One way to approach this issue is to recognize that sextortion, at the extreme end of the cyberbullying spectrum, is a way of using electronic information to force someone to do something they don't want to do. The place to start, then, is with an effective anti-cyberbullying program, both at home and in the schools. The initial focus of the program, for younger children, should be teaching them what is and is not appropriate to photograph and/or distribute on the Internet. As kids get older, the emphasis should shift to specific standards for online interaction; this is essentially the well-established anti-cyberbullying curriculum, enhanced by adding a privacy element. By middle school, kids should know (or at least it should be taught in the school's curriculum) that they shouldn't violate their friends' or classmates' privacy by taking their photographs, downloading their images, or publishing their information (including photos) without permission. As kids enter high school, they're old enough to begin learning the laws governing misuse of private information, including sextortion, and the criminal consequences for breaking those laws.

Realistically, though, this is not one of the cybertraps where mere communication is going to be enormously effective. Even under the best of circumstances, teens are often reluctant to talk, and there is little likelihood of a teen voluntarily telling his parents about his burgeoning career as a sextortionist. Nor is casual surveillance likely to do much good; again, if your child is committing this crime, he or she will go to great lengths to keep you from seeing evidence of it on the computer.

Surveillance software is not the be-all-end-all, nor is it a substitute for engaged parenting. Still, there's no question that, in some instances, surveillance software will alert you to the possibility that a crime, and potentially a serious crime, is being committed by your child. You may choose to inform your child that surveillance software has been installed, for there is a chance mere knowledge of the software's presence will be deterrent enough. Keep in mind, however, that your child might not view it as a warning so much as a challenge—a puzzle to be solved.

There are a couple different ways in which to understand surveillance of your child's online activity: it's either an insurance policy against accidental misconduct by a child still learning the rules of lawful behavior or an unfortunate necessity for a child you fear hasn't fully absorbed those rules and the serious consequences for breaking them. Many parents may be uncomfortable with the first idea and, understandably, even more reluctant to concede the second. Surveillance is an uncomfortable word to use when describing the supervision parents should provide their children, but these are technological times and sometimes a technological solution is required.

Chapter Twelve
The Costs of Cybertraps

As a parent, I know all too well how difficult it can be to get children to think more than a few minutes ahead. Lectures that include warnings about future consequences—no matter how dire—tend to fall on deaf ears. It is unrealistic to think that a simple conversation or two will flip a switch and our children's online behavior will miraculously change. But it's important to at least try to impress upon children that their online activity can produce real consequences—not just for them, but for their parents as well. My hope is that this chapter will underscore just how serious those consequences can be.

Derailing Your Child's Academic and Employment Future
1. No More Secrets?
Falling into one of the cybertraps discussed in this book can, first and foremost, put your child at a serious competitive disadvantage when applying for college admission, scholarships, graduate school, and employment. Our children are growing up in a hyper-competitive world, where even the slightest misstep can spell the difference between admission and rejection at a desired college, or between a second interview and a "thanks, but no thanks."

It should come as no great shock that colleges are scrutinizing academic records with increasing vigor, but it may surprise you to learn that admissions committees are increasingly searching the Web for information about applicants. Last year, a study by the National Association for College Admission Counseling (NACAC) found that 25 percent of colleges use "Web search or social networking technology to locate information about prospective students." At the corporate level, too, such searches are common. Three-fourths of recruiters say they use Web search tools and social networking sites to screen prospective employees, and more than a third say they've deep-sixed an applicant based on information they've discovered online.

If your child has been disciplined for an online transgression during middle or high school, exactly how much information will be available to an admissions committee or job recruiter will depend primarily on the policies of your child's school. However, thanks to Google, schools are no longer the sole gatekeepers of information about such incidents. A Google search could turn up news reports that mention your child's name, or documents related to a lawsuit in which your child was involved, even if he or she was a minor at the time. Your child's name might also show up on blogs or comments on social networking sites.

It's smart, then, to periodically Google your child's name to see what information might be publicly available. Even if your child hasn't committed a cyber transgression, there's still the possibility of someone posting inaccurate or even defamatory information about him or her. It's in your best interest to find that information before an admissions officer or recruiter does.

2. The Cost of a Criminal Record

Many of the cybertraps discussed in this book carry the risk of burdening your child with a criminal record. The good news is that if your child can be tried in juvenile court, his or her record will often be sealed, and the court will likely offer your child a clean slate after a period of good behavior. If criminal charges are filed against your child, you'll need to consult with a good defense attorney to review all the options. And keep in mind: virtually all college and employment applications ask not only about an applicant's criminal record, but whether he or she has faced any disciplinary proceedings. For instance, the common application used by many U.S. colleges and universities poses two relevant questions:

- Have you ever been found responsible for a disciplinary violation at an educational institution you have attended from ninth grade (or the international equivalent) forward, whether related to academic misconduct or behavioral misconduct, that resulted in your probation, suspension, removal, dismissal, or expulsion from the institution?
- Have you ever been convicted of a misdemeanor, felony, or other crime?

If an applicant answers yes to either question, he or she is required, on a separate sheet, to provide the approximate date for each incident as well as an explanation of the circumstances. That's one college essay your child certainly does not want to write.

There will be those who are tempted to answer falsely, particularly if their offense records are under seal in juvenile court, but keep in mind that a number of court decisions have upheld the right of a college or university to cancel or revoke a student's acceptance upon learning that he or she lied on an admissions application. And the chances of getting caught are greater than you might think, in part because of the increasing "Googlization" of the world, and in part because the cost of an old-fashioned background check continues to drop.

3. No Redemption: Sex Offender Registries

A criminal record can pose big risks to your child's educational or employment prospects, but inclusion on a sex-offender registry is even worse. Congress in 1994 mandated the creation of such registries at the state level (as with enforcement of the drinking age, states were threatened with a loss of federal funds if they failed to comply). In the years since, lawmakers have periodically expanded the list of crimes that can land someone on a sex-offender registry. Most recently, in the Adam Walsh Act of 2006, Congress created a national sex offender registry and created a three-tier classification system for sex offenders based on the severity of the underlying crime. Among other things, Congress also imposed a requirement for each state and territory to create and maintain a similar registry, and for information to be shared among the national and jurisdictional registries.

According to Human Rights Watch, an organization that has been working to change the rote application of sex-offender statutes to children—particularly in sexting cases—both federal and state laws currently require juveniles and other offenders to remain on sex-offender registries far past the time when they realistically pose a threat. Tier I offenders (those convicted of the lowest-level offenses) must register for fifteen years, Tier II offenders for twenty-five years, and Tier III offenders for life. Moreover, HRW points out, seventeen states require all sex offenders—no matter how minor and non-coercive their offense—to register for their entire lives. Examples include seventeen-year-olds who engage in consensual sex (over 55 percent of them do according to one report) or a sixteen-year-old who sexts a photo of herself.

In two states, Alabama and South Carolina, there's no mechanism available for a former offender to petition for removal from the list. In the other thirty-three states, the length of registration varies depending on the severity of the crime, although a repeat offender will be faced with lifetime registration. Since each state has its own registration and reporting requirements, it can be a real challenge for a registered sex offender to avoid committing another crime by failing to properly notify authorities of his or her location when changing schools or jobs.

While it's not impossible for a registered sex offender to get into college or land a job, it's dramatically more difficult. Schools and employers tend to take a long, hard look at applicants who have sex-crime convictions in their backgrounds. Some schools or employers may take a lenient view of a non-coercive sex offense like sexting, but other cybertraps discussed in this book—cyberstalking and child pornography, for instance—will not only require registration but will permanently scar your child's educational and employment prospects.

Congress in 2000 adopted the Campus Sex Crimes Prevention Act, which requires colleges and universities to maintain information regarding all sex offenders who are "employed, enrolled, or volunteering" at the institution. Schools must also periodically inform the campus community where such information can be viewed.

The Long Arm of the Law: Criminal Investigations and Computer Forensics

There's one word of caution you can give to your child that may make a real, immediate impression: if state or federal law enforcement has reasonable grounds for believing he or she has committed a cybercrime—downloaded child pornography, for instance, or cyberstalked someone, or engaged in hacking—they will come and search the entire house for electronic devices and data-storage tools. Anything that may have been used to perpetrate the alleged crime will be taken away, and the person responsible will be arrested.

Many of the cybertraps discussed in this book can result in a police search of your home. It is worth noting, however, that the cybertrap most likely to trigger such a search is child pornography, either as a result of P2P monitoring or a sexting investigation. Both state and federal law-enforcement agencies have devoted enormous resources over the last decade to fighting the scourge of child pornography. It's an uphill battle, to say the least. Since the introduction of digital technology, the production of new child pornography has exploded, and U.S. law enforcement struggles to deal with the large quantities of materials produced beyond the nation's borders and distributed online. But there should be no illusions whatsoever about the ability of federal and state investigators to track domestic online behavior to its source. The rise in the number of child-pornography investigations and arrests over the past decade is impressive, and there's one key reason for investigators' successes: a seemingly innocuous cluster of four numbers assigned to each and every device that connects to the Internet, a string of numbers that follows us everywhere we go in the wilds of cyberspace.

1. The Internet Protocol Address

To better understand the enormous pool of data available to law enforcement during a cyber-crime investigation, it's important to understand at least a little about how data is transmitted across the Internet.

First, all information that travels across the Web is broken up into one or more packets, each of which is transmitted independently from the sending computer to the receiving computer, then recombined at the destination to provide a coherent message. The structure of these data packets is governed by a set of rules called the Internet Protocol (IP). Among other things, the IP requires that each packet contain a source IP address and a destination IP address.

Currently, an IP address consists of a cluster of four decimal numbers, ranging from zero to 255. If you type "microsoft.com" into your Web browser, behind the scenes a request is sent to a server located at the IP address 65.55.12.249, which then sends the requested website back to your computer. (You can test this by typing that IP address into your browser instead of "microsoft.com.")

The four-decimal clusters, known as IPv4 addresses, have been in use since the earliest days of the Internet, and when the scheme was first developed it was assumed there would be enough addresses (roughly four billion) to last forever. But that was before the invention of the personal computer and Internet-capable mobile devices. So many electronic devices now connect to the Internet worldwide that the pool of available IPv4 addresses is due to run out sometime in 2011 or 2012. Fortunately, there's a fix in the works: a new IP-address scheme called IPv6, which increases, astronomically, the number of available addresses. There's a downside, however: while we'll be able to connect more devices to the Internet, it will also be possible to track more of what we do with those devices, since virtually everything will have its own identifying address.

2. Real-Time Monitoring of Peer-to-Peer Activity

The best example of how IP addresses are used by law enforcement arises in the context of peer-to-peer software. In order for computer users to share information across a P2P network, the software needs to know the IP addresses of both the computer searching for content and the computer offering to share it. P2P programs use the IP addresses to properly route files from one computer to another; they also include that information in the search results displayed on screen.

Not surprisingly, law enforcement has caught on to the fact that P2P software is used for distributing child pornography, and they use enhanced versions of the software to run their own searches for illegal content. When they find a computer offering child pornography, they record the IP address of that computer in a special database. An organization called the National Center for Missing and Exploited Children (NCMEC) maintains a growing database of electronic images and movies depicting victims of child exploitation, i.e., images where the identity of the pictured child has been established at trial. The entries in the database are described as Known Files, and law enforcement can compare images they find in their P2P searches to the Known Files database maintained by NCMEC.

Numerous law-enforcement agencies around the country are now using specialized software that can identify Known Files traveling across P2P networks in real time, and display the location of the computers downloading those files on a Google map that shows the physical address paired with the IP address of the receiving computer.

3. Search and Seizure

So how do law enforcement agents turn an IP address into a knock on your door? It's not particularly complicated. Since each data packet that travels across the Internet contains a sending address and a receiving address, it's possible for every computer that handles the data to maintain a record of which computers are communicating with each other.

There are numerous websites that enable anyone to look up the name of an organization using the IP address assigned to it. In the case of most home computer users, the IP address will be part of a wide range of addresses assigned to an Internet Service Provider like Comcast, Earthlink, or America Online. Investigators will send a request for information to the appropriate department, listing the specific IP address in question and the date and time of the activity they are investigating.

Kids tend to have a hard time understanding just how many organizations can gather information about their online activity. Even a relatively simple text message sent from one mobile phone to another must travel through various cell towers and a wireless carrier's computer network, where a record is created listing the sender's phone number, the receiver's phone number, and, in some instances, the message itself (many mobile-phone companies, in fact, will send copies of teen text messages to parents—but if your child sends thousands of texts a month, you might want to think carefully before signing up).

When it comes to social networking, the number of potential intermediaries is even higher. Home Internet access is typically provided by an ISP like Verizon or Comcast, and all ISPs record and retain basic information about their subscribers' Internet activity—their IP addresses, the time each Internet sessions starts and ends, and sometimes information about the online activity itself.

Whether or not the ISP will be able to tell investigators exactly who was using a particular IP address at a given date and time will depend largely on how much time has passed. ISPs don't maintain those records indefinitely (the average length of retention is from three to six months). Some law enforcement officials (most notably former U.S. Attorney General Alberto Gonzales) have advocated for much longer data-retention periods, but those proposals have met with stiff resistance due to privacy concerns.

In general, however, law enforcement is able to move quickly enough to request IP address data while the ISP still has it. The ISP will search its records and provide investigators with the name and physical address of the person listed on the account. An investigator will then take that information and apply for a search warrant, which describes his or her reasons for believing that evidence of a crime will be found at the address.

4. Computer Forensics

In most cases, the first inkling you'll get that your child has been targeted by law enforcement as a child-pornography suspect will be when a police officer or sheriff shows up at your door with a search warrant for the premises. If you're presented with a search warrant, call your attorney, and read the warrant carefully. You can't prevent the execution of a valid search warrant, but you are entitled to make sure police comply with its limitations, and your attorney can help you make sure the warrant is executed properly.

A valid search warrant should specify not only the premises to be searched, but should provide a detailed list of items and materials for which police are searching. In the case of a computer-related crime like child pornography, the warrant will typically have fairly comprehensive language authorizing the police to search for any possible storage location. As an example, here's sample language from a search warrant in a case on which I recently worked:

> Computer hardware to include any and all computer equipment used to collect, analyze, create, display, convert, store, conceal, or transmit electronic, magnetic, optical, or similar computer impulses or data. Hardware includes (but is not limited to) any data-processing devices (such as central processing units, personal computers to include "laptop" or "notebook" or "pocket" computers); internal and peripheral storage devices (such as fixed disks, external hard disks, floppy disk drives and diskettes, tape drives and tapes, optical storage devices, and other electronic media).

The warrant went on to specify another ten paragraphs' worth of items police intended to search. Depending on the jurisdiction in which you live and the training of the officers who applied for the search warrant, the language will vary.

If police locate items within the scope of the warrant, they will first do a field review of each item to see if it contains child pornography. If it does not, typically the item won't be seized, unless it contains some other potentially relevant evidence, such as e-mails or passwords. If the field review *does* detect child pornography, or if police are unable to make an initial determination, the item will be taken back to a lab for a more thorough forensics review. In such cases you should be given a signed inventory listing each item the police seize from your home.

At the lab, a trained computer technician will make an exact duplicate of each potential evidence item (hard drive, USB stick, CD-ROM, etc.) and use specialized software to search for contraband. What many people don't understand is that computers do a poor job of actually wiping out files deleted from a computer, and in a significant number of cases large numbers of deleted files can be recovered from a hard drive or external drive.

In addition, computer forensics software does an excellent job analyzing the files stored in your computer's Internet browser cache. Those files can contain full copies of websites that have been viewed, as well as individual images, banner advertisements (which can be quite explicit), and so on. In many cases, the searches entered into sites like Google and Yahoo are also tucked away in the cache. All this makes it fairly easy for a computer forensics expert to put together a thorough profile of how a given desktop, laptop, or even mobile phone has been used over a period of time.

In short, if your child's online activity has generated enough law-enforcement interest to lead to the issuance and execution of a search warrant, the odds are good police will find something to support their interest. Once they've done so, the question will be what charges your child could face, what defenses can be raised, and how expensive it will be hire an attorney to handle his or her defense.

The Financial Impact

1. Legal and Expert Witness Fees

For some cybertraps, the cost of extracting your child will be relatively minimal. Transgressions like plagiarism or low-level cyberbullying rarely require the assistance of an attorney. But if your child's online behavior sparks a lawsuit, or if he or she is being prosecuted, then the services of a good lawyer are essential.

How expensive will those services be? That depends on any number of factors. In civil litigation, the legal costs are driven in large part by the complexity of the issues involved, the number of parties, and the potential damages. In criminal matters, the number and severity of the charges will largely determine the cost. In both cases, your child's attorney may recommend hiring an independent computer forensics expert to evaluate the evidence, help the attorney understand the facts and formulate strategies, and, if necessary, testify at trial.

All cases are unique, so there's no surefire way to predict how expensive defending your child in a civil or criminal matter would be. Suffice it to say, however, that the cost of even a routine civil or criminal case—between attorney and expert-witness fees—can easily reach into the tens of thousands of dollars.

2. Civil Damages and Criminal Fines

There are two other types of financial expense associated with the cybertraps discussed in this book. The first: damages that can be awarded by a judge or jury if your child loses a civil case for copyright infringement, invasion of privacy, defamation, or a similar tort. Since damage awards are so fact-specific, there's no basis for estimating what might be assessed in a given case. It's worth noting, however, that to the extent the Internet speeds up copyright theft and expands the potential audience for defamatory comments, it correspondingly increases the potential damages. As Louis Brandeis and Samuel Warren warned in "The Right to Privacy," "numerous mechanical devices threaten to make good the prediction that 'what is whispered in the closet shall be proclaimed from the house-tops.'" The further and louder these statements carry, the more potential damage they cause.

Those facing charges for criminal cybertraps may also find their wallets (or their parents' wallets) significantly lighter. Most criminal statutes provide for a fine in addition to—or instead of—a prison sentence. In addition, many statutes now also provide for restitution to a victim for damages he or she may have suffered as a result of the criminal act. Your child's defense lawyer can explain those possibilities to you if the need arises.

Part Three: The Solutions

Chapter Thirteen
DON'T Stop Educating Yourself

If there's one aspect of technology that tends to freak parents out, it's the pace at which it changes. Few of us have the free time required to keep up with the constantly expanding universe of electronic devices, applications, and Web services. And teens—given their natural desires to carve out private spaces in which to experiment and grow—tend to actively resist parental supervision of their activities, electronic or otherwise.

Fortunately, you don't have to become a technological guru or CIA-level surveillance expert in order to better protect your children as they grow up in this increasingly wired (and wireless) world. With a fairly small investment of time, and by asking the right questions, you can gain enough knowledge to steer your children past the cybertraps discussed in this book.

Talk to Your Child

I said this at the beginning of the book, but it bears repeating: one of your best sources for information about new technology is your own child. Let's say, for instance, that your thirteen-year-old daughter has been pleading for some new piece of electronic gadgetry. Before agreeing to purchase it for her, ask her to explain exactly what it does, and how she plans to use it. Equally important: ask which of her friends own similar technology, and what *they* do with it. For instance: is the technology used to talk, text, send photos, or even some combination of all three? Make it clear that you won't even consider buying your daughter this new device until she first has a frank discussion with you about its potential uses (and abuses).

Odds are good that whatever the particular device, it can be used to communicate with others. If that's the case, it's a good opportunity to ask a few questions. Who does your daughter plan to talk to, or text? What does she plan to say (or show)? Does she plan to add to the device's capabilities by downloading apps or additional programs?

Your child may not be aware of everything a device can do, but rest assured it won't take her long to figure it out. Try to get ahead of the curve by researching new purchases at the time they're made, and don't forget to check back with your child from time to time to ask how she's actually using the device.

Talk to Friends, Teachers, and School Administrators

When your child starts expressing interest in a new technology or a specific brand of electronic gadget, take the opportunity to begin talking with other adults about their experiences with the product. Given the herd mentality of most preteens and teens, the odds are overwhelming that someone in your family's social circle has experience with the technology in question and can give you pointers on its capabilities, as well as suggest issues you'll want to discuss with your child.

It can be particularly useful to ask other parents what household rules they've established regarding the use of a particular device. Their experience can help you formulate your own guidelines and usage boundaries. And that kind of networking between parents will help you present a united front to your children.

It's also a good idea to check in with your child's teachers, as well as school administrators, to ask about their experiences with the device in question. Ask if they're aware of any discipline problems that have arisen, or particular concerns they might have about how students are using a given technology. Also be sure to ask about specific school policies concerning the device, or personal electronics more generally.

Visit a Few Key Websites

There are two kinds of websites that will help keep you up to speed on the technology your child uses, or craves, and the potential cybertraps that might be lurking around the corner.

First, there are a number of websites that track and review the latest cool gadgets. Some specialize in electronics themselves: Engadget, Gizmodo, and Coolest-Gadgets, for instance. Others offer more general tech coverage: CNET, PC World, PC Magazine, Macworld, MacRumors, and AppleInsider. You could easily spend hours, or even days, surfing these sites. Of course, unless you're a total tech monkey, reading all these tech sites can become a little eye-glazing, so you might just ask your child to explain how he heard about the gadget he craves. Depending on his age, he'll either be constantly reading tech sites himself—a chance for you to ask which ones—or he will have heard about the gadget from friends. In either case, your conversations about where the buzz comes from will give you ample opportunity to warn your child about the possible consequences of misuse.

A second category of sites is made up of those aimed at helping you understand the legal ramifications of the cybertraps that can capture your child. The website for this book (www.cybertrapsfortheyoung.com) is a good place to start, since it's been specifically designed as a repository of information about the legal risks children face online.

You can also glean a good bit of information by perusing major news outlets and looking for stories about teen perpetrators. All the major news sites—my favorites are *Wired,* the *New York Times,* the *Washington Post,* the *Boston Globe,* CNN,

HuffingtonPost, and *ars technica*—carry stories about teens and preteens who have committed cybercrimes, but the coverage is sporadic and locating relevant stories can be time consuming.

As you might expect, there are a variety of websites aimed at keeping your child safe online, though they tend to be written from the perspective of the child as victim, rather than as potential perpetrator. Nonetheless, those sites can be helpful, and they're particularly good sources of information about surveillance options. Good places to start include attorney Parry Aftab's WiredKids.org, the Google Family Safety Center, the Parents & Guardians page run by the National Center for Missing and Exploited Children, and GetNetWise.com. Links to those sites and others can be found on this book's website.

Online Search Is Your Friend

If you're trying to keep pace with your child's technological savvy, don't overlook the value of online search. Google is the largest search engine, but you can also get good results with its leading competitors, including Microsoft's Bing, Yahoo! Search, and Ask.com.

There are a couple different approaches to take with these search sites. For starters, if you're curious about a specific issue, you can use a quick word search— "teen cyberbullying prosecution," for instance—to pull up a huge amount of information in a matter of seconds. You can narrow down that information as you see fit—by state, city, type of device, etc.

Another powerful tool is the Google News site (news. google.com), which I use frequently. Here you can use the same types of keywords and phrases you'd plug into a regular search box, but the results will be limited to news items. For instance, the phrase "teen prosecution hacking" turns up a long list of articles about the Dutch teen discussed in chapter 9. Google also allows you to receive e-mail alerts for a particular search; just click on the link that appears at the bottom of any news search, and new articles will be sent to your inbox as they crop up.

It won't take long to gather enough information to talk sensibly and effectively to your child or, at the very least, to start a conversation with some probing questions. But it's important to be clear on one final point: if you think your child has been snagged in any of the cybertraps I've described, or if you're contacted by your child's school or the police, it's time to invest serious and focused effort in learning what your child has done, how he or she has done it, and the potential legal consequences. If you don't feel that you have the technical skills to conduct a thorough computer investigation yourself, hire someone who does—either a local computer technician or, if necessary, a computer forensics expert. On the legal front, make sure that the criminal lawyer you hire is familiar with computer technology, or regularly consults with an expert who is. And continue to educate yourself: the more you understand about the technology in question, the better you'll be able to communicate about it with others, and the more effectively you'll be able to advocate on your child's behalf.

Chapter Fourteen
DO Understand the Impact of Technology on Your Child

It's tempting, after learning about the many ways in which our children can get themselves into serious trouble online, to simply throw open a window and start heaving their electronics out. And even if our kids never find themselves in legal hot water, their obsession with technology, their tendency to cocoon themselves in headphones, and their incessant texting may all seem like equally good reasons to turn yourself into a neo-Luddite. But, of course, it's not that simple. An important part of grappling with these issues is understanding the ways, both positive and negative, in which technology is affecting the development of our children.

Children are a barometer of the future, and their some-times-fanatical use of electronics is a preview of the changes that will reshape our entire society in the coming years. Many of the technical and social networking skills your children are developing will become critical components of their working lives. And researchers are now learning that electronic devices—and the new forms of communication that go with them—are an increasingly important part of a child's social and even neurological development. In a 2008 survey of two thousand teens by CTIA-The Wireless Association, 57 percent reported that their mobile phone was "the key" to their social life, while 47 percent said—with typical teen hyperbole—that their social lives would "end or be worsened" if they could no longer text. It wouldn't surprise me if both percentages were now even higher.

Kids have plenty of legitimate reasons for liking technology. For instance, according to the CTIA survey, four of every five teens—roughly the same number who routinely use a mobile phone to get a ride—say carrying the phone provides them with a sense of security. Half say they use their mobile phones to locate important information, and a third have used it to help someone who is in trouble.

Children, Technology, and Media Consumption

Security and convenience of transportation are valid reasons for having a phone, and the ones most likely to be cited by your child (if he or she is savvy) when petitioning for a new or upgraded model. But that's not the full story. Parents also need to understand the impact technology can have on a child's social, intellectual, and even physical development.

In April 2010, the research and consulting firm OTX surveyed six hundred teens between the ages of thirteen and seventeen about their experience with social networking and other online communities. The survey results offer useful insights into how teens spend their time online:

- Teens are spending just under two-and-a-half hours per day on the Internet.
- Of that time, an hour and fifty minutes is spent on social-networking sites.
- The top-ranked social-networking sites are: Facebook (69 percent of teens have a profile and use it), YouTube (64 percent), MySpace (41 percent), Twitter (20 percent), Windows Live Spaces (16 percent), Pandora (15 percent), Gaia Online (14 percent), and Club Penguin (13 percent).

- In an indication of how fragmented the social networking scene is, teens listed nearly twenty other sites on which they maintained profiles, ranging from the relatively well known (LinkedIn, FourSquare, Yelp) to the obscure (Netfog, Looklet, Raptr).

Early in 2010, the Kaiser Family Foundation released the results of its third national survey on media use among children aged eight to eighteen. Not surprisingly, the foundation found that child media consumption had risen dramatically over the prior five years, from just under six-and-a-half hours per day to a staggering daily intake of seven-and-a-half hours. Moreover, the foundation said, because children typically multitask among different media sources, they actually consume nearly eleven hours' worth of media each day.

The primary factor behind the increase is the popularity of mobile multimedia devices that enable children to access the Internet, watch TV, and stream movies on the go. Interestingly, in a harbinger of the future for television networks, kids in 2010 spent nearly half an hour less watching regularly scheduled TV programming than they did in 2004, the first decline recorded by Kaiser. However, due to the myriad ways in which kids can watch "television" programs, the total amount of time spent on TV media actually increased, from just under four hours each day to roughly four-and-a-half hours.

Technology and a Child's Social Development

The Kaiser Foundation also asked about texting; teens said they spent about an hour and a half each day sending and receiving messages. Significantly, the foundation didn't include texting in its calculation of total media use among teens: that hour and a half comes on top of the more than seven hours kids spend with other forms of media. Is it any wonder that our students are lagging behind other industrialized nations in academics but leading them in childhood obesity?

There's still room for debate, however, about the role that technology—particularly massive amounts of texting—plays in our children's development. Researchers at the Indiana University Department of Telecommunications, in a 2006 paper, sought answers to three key questions:

1. Are adolescents creating more, but weaker, ties using socially interactive technologies (SITs)?
2. To what extent do adolescents' SIT communication networks overlap with their real-world (offline) friendship networks?
3. Are SIT-based relationships more important for adolescents who have fewer offline peer ties?

The researchers concluded that the answer to the first question was no: teens were not creating larger numbers of weaker relationships through the use of SITs. In part, the study suggested, this could show that teens these days make fewer distinctions between online and offline relationships.

On the second question, research indicated that—at least in 2006—there wasn't much overlap between a child's offline and online friend networks. One would suspect this finding has changed, however, both with the increasing popularity of Facebook—which only opened its doors to high-school students in September 2005—and the sharp rise in the number of teens carrying mobile devices.

Lastly, the researchers concluded that at-risk adolescents were not, in fact, expanding their pool of relationships through socially interactive technologies. In fact, the researchers found that at-risk adolescents formed proportionately fewer SIT-based relationships or, in many cases, none at all.

Other studies have suggested that texting, far from ruining your child's ability to use "brother" instead of "bro," actually improves overall proficiency with language. Similarly, texting is proving to be a powerful tool for student/teacher interaction in the classroom, in no small part because kids are so attuned to it.

Rewiring the Brain

While there may be benefits to all this new technology, the consensus seems to be that it's having a mostly deleterious effect on our children's learning habits and face-to-face social skills.

Scientists are beginning to discover, for instance, that the Internet actually has a discernable effect on our brain structure, and the size of the effect is directly related to the amount of time a person spends online. In February 2009, Baroness Susan Greenfield—a noted British researcher of brain physiology and, at the time, director of the Royal Institution of Great Britain—suggested that social networking sites might be "infantilizing the mid-21st century brain." In particular, she cautioned that incessant social networking increased the likelihood of "short attention spans, sensationalism, inability to empathize, and a shaky sense of identity."

Meanwhile, Sherry Turkle, a professor of social science and technology at MIT, argues that texting has become an addiction for teens, not only causing them to persistently seek the rush of validation from their friends, but also damaging their ability to experience intimacy with other human beings.

Turkle, like other researchers, also worries that texting can cause emotional and physical harm. Some pediatricians, Turkle told National Public Radio, report patients with difficulty sleeping and pain in their hands (particularly their thumbs) from the repetitive motions of typing on miniature keyboards. In fact, a 2010 study at the Case Western Reserve School of Medicine found disturbing correlations between the practice of hyper-texting—which they defined as sending more than 120 texts per day—and a variety of negative health consequences.

According to the Case Western study, hyper-texters were 40 percent more likely to have tried smoking, twice as likely to have tried alcohol, 43 percent more likely to binge drink, 41 percent more likely to have used illicit drugs, 55 percent more likely to have been in a physical fight, nearly four times as likely to have had sex, and 90 percent more likely to have had four or more sexual partners.

There's one last obvious concern: sleep deprivation, either from staying up too late texting and surfing on mobile devices, or from leaving the device on and waking up repeatedly during the night to read and respond to texts. In November 2010, doctors at the Sleep Disorder Center at JFK Medical Center in Edison, New Jersey, told colleagues at an annual conference that nearly four out of five teens who use electronics at bedtime report "persistent problems" falling asleep. "Children who engage in pre-bedtime use of technology have a high rate of daytime problems, which can include attention-deficit/hyperactivity disorder, anxiety, depression, and learning difficulties," said Peter Polos, the study's lead author. "This is in addition to nighttime problems, such as excessive movements, insomnia, and leg pain."

All these studies illustrate the essential conflict between teens and parents when it comes to technology. For teens, technology is an integral part of their lives and a valuable adjunct to their social development. For parents, the challenge is how best to honor those legitimate feelings while minimizing the potential harms. Some house rules can help.

Chapter Fifteen
DON'T Let Computers and Other Technology Out of Common Spaces

In 2000, when I was traveling around the country on my book tour for *Obscene Profits* (ah, those were the days in publishing!), I had a chance to be a guest on a number of call-in radio shows. The Internet, at least in the eyes of the public, was still a relatively new phenomenon, and a lot of parents were just beginning to realize how much adult content was available to their children online. When I asked callers about the location of their home computer (at that time, most households had only one), nine times out of ten the answer was, "In my child's bedroom."

I tried to explain, as politely as I could, that leaving an Internet-connected computer in the bedroom of a child wasn't so different than putting the family liquor cabinet next to a teenager's bed. Given the naturally inquisitive tendencies of teens and preteens, to say nothing of their fondness for staying up far past adult bedtimes, unfettered access to the computer meant that exploration of the Web's wilder side was virtually inevitable.

The suggestion that parents put the household computer in a common space made sense, as well, as the household computer was really the only available portal to the Web. Of course, even then it wasn't a surefire means of preventing kids from accessing inappropriate content, particularly since so many families included two working parents, sometimes with multiple jobs. It's hard to supervise what a child is doing online, after all, when you're working a second or third shift. But it was a simple step that sent a clear message to children that their parents cared about what they were doing online.

Today, supervision of teen and preteen online activity is even more difficult than it was ten years ago. The cost of computer technology has fallen steadily, to the point where it's not uncommon for each person in a household to have his or her own computer, most likely a laptop. A significant number of children now have one or more mobile devices that enable them to access the Web and communicate with their friends, and the speed of home Internet access has increased phenomenally in the past decade. In 1999, for instance, a state-of-the art Dell laptop came equipped with an internal 56K modem that needed a telephone line to connect to the Internet. It's not that kids couldn't get caught in cybertraps at 56K, just that it took a lot longer. These days, lightning-quick DSL and cable Internet connections make it possible for kids to watch streaming video—not all of it kid-friendly, by any means—and download songs, albums, or even movies within seconds.

Does the Common Space Rule Still Apply?

So the question becomes, "Does the common-space rule still work?" And the answer is yes, at least until your child hits middle school, when more invasive monitoring may be required. When children are in elementary school, they're somewhat less likely to want or need their own computer, so it's very easy to require them to use a common family computer, where you can see what they're doing, answer their questions, and educate them about online safety. It's also a good opportunity, over a period of months and years, to remind them that supervision (they may call it surveillance) is your ongoing responsibility, and that electronic privacy is not a right for children, but something they can begin to earn as they demonstrate sufficient responsibility and maturity.

Of course there may be periods when even a young child is in someone else's care. If you're not able to watch your child full time because of work, or some other obligation, make sure your child's caregiver understands the boundaries you'd like enforced when it comes to computers and other electronics. Assuming that your caregiver is responsible, there's probably no need for surveillance software on the computer.

Managing online access gets significantly more complicated as your child heads into middle school. He or she will undoubtedly want to spend more time communicating with friends online, either through instant messaging or various social networks, and will begin to push for his or her own laptop or smartphone. Depending on your child's school and level of homework, that may seem like a good idea, particularly if conflicts have arisen over access to your—or the family's—computer. It's also likely that your child will start spending more time at the homes of friends, and that exposure to different household rules will raise additional questions about supervision.

At the same time, your child's demand for personal privacy will intensify, as will his or her resistance to using the computer only in a common space. It's a difficult balance to strike. On the one hand, it's natural and appropriate for children to want more privacy. On the other hand, that developmental stage coincides directly with an increased risk for many, if not most, of the cybertraps discussed in this book.

There's no easy or pleasant resolution to the conflict. If you insist that your child continue to use a computer in a common space, the chief downside will be a perpetually annoyed adolescent. On the other hand, if you decide it makes sense for your child to have a laptop, or desktop computer stationed in his or her room (allowing a quieter environment for schoolwork), it's time to think seriously about investing in surveillance software. That way you can make sure your child is navigating the social challenges of middle school without resorting to inappropriate or illegal online behavior.

How long and how intensely you monitor your child's online activity will depend on the personality of your child, the extent to which he or she has absorbed lessons about ethical online conduct, and your level of comfort about his or her maturity. Provided that your child has given you full friend status on social networks, and assuming that periodic spot checks of his or her online activity don't raise concerns, you may feel comfortable easing up on the level of supervision as your child progresses through high school.

The Family that Charges Together

Internet-capable mobile devices raise their own special issues. Smartphones and tablets have virtually all the capabilities of full-fledged computers, and their portability means our children can take them to places where we can't easily monitor their activity. Mobile devices can also be taken more easily to bed, where they can keep children up and disrupt their sleep patterns. It's also more difficult to monitor your child's Web or communication activity on a mobile device than it is on a computer, although a number of companies have started offering surveillance software tailored to such devices.

I've found that children (and plenty of adults) are more attached to their mobile devices than they are to their laptops. There's something about the portability and the sense of connectedness that makes it hard for people to set them aside for any length of time. (Believe me, I speak from experience: I lost my phone in Chicago a year ago and I was completely out of sorts until I replaced it.)

However, we all need to carve out a time to put our electronics aside, and that's particularly important for children.

One solution is to designate a central location in the home for a familial charging station, where at a certain hour all mobile devices go to rest up for the next day's labors. If you think your child might be unable to resist an incoming text in the middle of the night, make sure the phone's turned off and charging in your bedroom. It makes sense that your sixteen-year-old would be allowed to stay up and communicate later than your twelve-year-old, but for each there should be a regularly scheduled time when the digital devices are put away, the electrons stop flowing, and it's time to rest. Or they can do what their parents used to do: sneak a flashlight under the covers and read a book.

Chapter Sixteen
DO Install Surveillance Software and Conduct Inspections of Mobile Devices

For Children, Privacy Is a Privilege, Not a Right

In many ways, this is one of the more difficult chapters to write, since I believe so passionately in the right to privacy for individuals. At the same time, I'm willing to accept that while certain constitutional rights are absolute regardless of a person's age, others do not come into full effect until a person is legally an adult. The right to privacy, I believe, is one of those rights.

There are nuances to this position, of course. I think children should have a greater ability to assert their constitutional rights, such as privacy or free speech, in public spaces, or in schools or libraries, than they do at home. Even within the home, there are some activities that are intrinsically more private (bathing or dressing, for instance) than others. But at the end of the day, children are children, not adults, and they often don't fully understand the potential consequences of abusing the rights that will be fully granted to them as adult citizens.

This is far from a novel concept when it comes to rearing children. Most parents are pretty careful about giving their kids unfettered access to sharp knives, alcohol, guns, or vehicles. As children mature, they're allowed to do more in the kitchen, and eventually graduate to unsupervised cutting and carving. In many households, children are taught to hunt, but not before they're given a thorough training in gun safety. The same, of course, is true for driving. The reason adults are leery of giving young children access to these potentially dangerous instruments is simple: kids lack the training and judgment to use them safely. Although some courts have disagreed, a mobile phone or laptop computer can also become a dangerous instrument in the hands of an immature or malicious child. The repercussions of misusing an electronic device may not be as immediate, nor as bloody, as the repercussions of misusing a handgun, but there's no question the effects can be painful and permanent.

The question to ask yourself when your child asks for electronic privacy is, "Privacy to do what?" If it's private space to explore his or her relationships with other kids, or to research topics he or she might not feel comfortable discussing with you, that's one thing. But if it's privacy to harass other students, practice identity theft or hacking, or sext a boyfriend or girlfriend, that's another story altogether. There's no reasonable expectation of privacy for criminal activity. Until you're confident that your child is not breaking the law, supervision and even surveillance of electronic devices is a parental responsibility and not—as you may be told—a villainous attempt to ruin your child's life.

The Value of Hands-On Inspections

When it comes to supervising your child's electronic activity, there's no substitute for a hands-on inspection of whatever devices your child uses, preferably with the child in question at your side. There are a number of advantages to this approach. First, it's a great opportunity to reverse roles and get some technology training from your child, who may be more adept than you are with electronic gadgetry. If you're having trouble understanding how a particular program or device works, or how to access certain information on the computer, ask your child to help.

Second, a hands-on inspection is the most efficient way to find out who your child is interacting with online, and to learn about the nature of those interactions. If you review your child's e-mails or text messages while he or she is sitting beside you, you can have an ongoing discussion about appropriate conversations and online activity. And by identifying your child's frequent contacts, you can compile a list of other families you'll want to bring into discussions about electronic usage. Not every family will have exactly the same standards, but the more alignment there is among parents of friends, the easier it will be for all of you to enforce the rules.

Third, periodic hands-on inspections will make it clear to our children that the continued use of electronic devices is a privilege, not a right. Inspections also demonstrate that we care about their development and want to protect them from unintended consequences as they mature. Some children may be acting intentionally when they get themselves ensnared in cybertraps, but many more stumble into them unwittingly—and even those who know what they're doing often don't consider the consequences of their actions. Having watched three boys transition through adolescence, with a fourth just beginning, I'm not sure it's even fair to talk about a "decision-making process" when it comes to teen behavior. Too often, the gap between thought and action is virtually nonexistent.

Surveillance Software: A Necessary Supplement

As I've noted frequently in this book, the realities of modern family life may prevent you from supervising or inspecting your child's use of electronic devices as frequently as you might like. If that's the case, it's appropriate to consider purchasing and installing surveillance software. What specific software you purchase, and how easy it will be to monitor your child's electronic activity, will depend largely on the type of device he or she is using.

Personal computers and laptops are the easiest to monitor. For the better part of a decade, software companies have been developing and refining monitoring tools for personal computers. A list of the most popular products can be found in appendix A. These software packages vary slightly, but nearly all of them perform several basic functions:

- Monitoring of Web activity, including e-mails, instant messaging, chats, Web browsing, and file downloads
- Remote access or Web access to data collected by the software
- The option of a stealth mode, which prevents your child from knowing he or she is being monitored
- The ability to log keystrokes
- Recording what programs are used and, in some cases, for how long
- The option to block access to particular websites based on keywords or a black list (particularly useful for younger children)
- The ability to filter or search Web activity based on pre-defined keywords
- The ability to record screen shots or videos of Web activity

Thanks to years of development, these software packages are quite effective in monitoring computer-based activity and alerting you to what is taking place. Many of them can even alert you by e-mail when restricted activities take place on the computer, or when particular keywords are used. But monitoring software also has its limitations, the most obvious of which is that it only works for the computer on which it's installed. If your child has ready access to more than one computer (at a friend's house, for instance), that activity won't be monitored. Similarly, if your child is accessing the Web or communicating with peers using a mobile device, separate surveillance software will be needed for each device.

If your child is particularly skilled with computers, he or she may be able to disable or otherwise tamper with the surveillance software. That would tip you off fairly quickly to your child's computing prowess, and it also should raise red flags about what he or she is doing online.

If you do decide to install surveillance software, there's one significant choice to make: whether to run the software openly on your child's computer, or to install it in stealth mode. For a number of reasons, I think open and obvious installation is the better approach.

For one, stealth mode really only operates in "stealth" until you see something questionable you feel you need to discuss with your child. You could tell your child you found the information from actually looking at the computer, but depending on your child's computer prowess and his or her perception of your computer skills, that may not be believable. Second, if your child is aware of surveillance software, that knowledge itself may prevent the kind of activity that could land someone in one of the more serious cybertraps. Of course, if your child is intent on doing something malicious or illegal, he or she will likely try to find a way to work around the software. However, for most children, the knowledge they're being watched will be a sufficient check on bad behavior. That's particularly true if you follow up with your child on early instances when the software shows he or she has crossed a line.

Third, surveillance of children is, at best, a necessary evil. While your children might not have the same rights to privacy that you enjoy as an adult, the less secrecy involved in your relationship the better. I've wrestled with this issue as a member of the Burlington School Board, where we've had to make decisions about installing surveillance cameras on school buses and in school buildings. As much as I dislike the practice, I grudgingly accept that it is many times an appropriate precaution in this day and age. Still, the necessity for surveillance is not a rationalization for spying, and we shouldn't start teaching our young children that spying is okay in all instances. Moreover, if you do use surveillance software, you should use it as an opportunity to teach your children about the ownership and control of data. Make clear to them how you will use the information you collect. Preventing possible criminal activity is an obvious use, but you might also use it to enforce household moral codes (for instance, restrictions on certain types of language or visiting adult websites). The open installation and operation of surveillance software provides a good opportunity to have a discussion with your child about what is acceptable or unacceptable behavior, and why.

Unfortunately, parents who want to add surveillance software for other portable electronics will discover that, in many cases, the technology is still in a transitional period. Thanks in large part to Apple's trailblazing products, the iPhone and the iPad, the thirty-year dominance of personal computers is rapidly coming to a close. Industry analysts expect tablet devices like the iPad to make significant inroads into PC sales in the coming year, and as those devices grow more sophisticated and more powerful the trend will only accelerate. Because these devices are so new, software manufacturers have not yet adapted their PC monitoring programs to work with them. (One company has begun advertising surveillance software for the iPad, but it's necessary to first "jailbreak" the device, i.e., alter its operation so that it will run non-Apple-approved software.) This poses a dilemma for parents: these devices are incredibly popular with kids—the Apple iPad was one of the top Christmas requests of 2010—but they're also difficult to monitor, and powerful enough to trigger most of the cybertraps in this book. Parents should think carefully about handing these tablets to children, particularly young children, without adequate supervision.

The surveillance options for mobile phones are at least a little more plentiful. Various companies have started to market software and services designed to monitor activity on smartphones. Most of these software packages are designed to operate in stealth mode—the products tend to be marketed to worried spouses more than concerned parents—so if you want to be open with your children about your surveillance you'll need to alert them to a software package's installation. Again, more details and a list of some of the leading surveillance options are included in appendix A.

Chapter Seventeen
DON'T Forget Video Game Consoles

As if mobile phones and tablet PCs didn't complicate matters enough, there's an entire class of Web-capable device for which there's absolutely no available surveillance software: video game consoles and handheld game consoles. A dozen years ago, that wouldn't have been a big deal, since the only things to monitor on a gaming console were the particular games being played, and for how long. But twenty-first century gaming consoles let your kids do a lot more than play games.

The Word of the Day Is Convergence
In the fall of 1999, Sony introduced its short-lived Dreamcast game console, the first to come with a built-in modem and support for online gaming. Although Sony stopped shipping Dreamcast consoles less than two years after their U.S. launch, the device was on the market long enough to prove the popularity of online gaming and the convenience—even with slow connection speeds—of downloadable content.

Over the last decade, the three major console manufacturers—Microsoft, Sony, and Nintendo—have each added numerous features to their specialized gaming devices that not only bring them closer in function to personal computers, but also increase the chances they'll wind up as accessories to cybercrimes. For instance, all three of the leading consoles now support keyboard input, social networking tools, and/or Internet access. Microsoft's Xbox 360, for instance, offers access to Windows Live Messenger, Facebook, and Twitter. The Sony Playstation 3 comes equipped with an Internet browser, video chat, and Facebook access. And Nintendo's Wii gives users access to the so-called "Internet Channel," which uses a version of the Opera browser specifically designed for navigation using a Wii controller, and gives users access to virtually any site on the Web.

Since gaming consoles run their own unique operating systems, there's no way to install standard Windows or Mac computer programs, which makes certain cybercrimes difficult or impossible to commit (serious hacking, for instance). However, your child could use the communication capabilities of any of these consoles to engage in cyberbullying or harassment. Similarly, the video-chat capabilities of the Playstation 3 could be used for sexting or sextortion. And of course there's ample opportunity for bullying within multi-player games themselves.

The lack of a standard operating system also means that existing surveillance software is useless for these consoles, and so far manufacturers haven't filled in the gap by creating their own surveillance tools. That doesn't mean, however, that the only way to supervise your child's activity on a gaming console is to sit in the room while he or she uses it. If your child is accessing social networking sites like Twitter or Facebook through a gaming console, you should be able to see that activity online, provided you've subscribed to your child's Twitter feed and mandated full friend status on Facebook. On the Playstation 3, you can see your child's Internet browsing history, just like on a computer (unless your child has gone to the trouble to erase it, which in itself may be a red flag). The Nintendo Wii, on the other hand, does not store information about what sites are visited using its Internet Channel.

If your child has one of the more advanced handheld gaming devices, like the Sony PlayStation Portable, Tiger's Gizmondo, or the Nintendo DS (with the appropriate Internet cartridge), he or she has the ability to browse the Internet and communicate with others using online social networking services. It's not as easy to do so on those devices as it is on a smart phone or a computer, but it's important to understand the convergence happening there. Not surprisingly, there's no software that enables parents to track the non-gaming activity occurring on a portable gaming device.

Video game consoles aren't the only devices on which this kind of convergence is occurring: thanks largely to Google, access to the online realm is also happening on your TV. To be fair, the idea of combining television and the Internet was first introduced nearly fifteen years ago by inventor and entrepreneur Steve Perlman, when he built an adapter for TV sets called Web TV, allowing users to surf the Internet and send e-mail. But operation of the device was clunky, particularly given the slow connection speeds and relatively low resolution of television screens at the time.

While Google TV is hardly perfect, it's come along at a time when Internet speeds are far faster and screens have much finer resolution. As a result, Google can credibly claim that its product is "an adventure where TV meets Web, apps, search, and the world's creativity." Google TV is an open-software platform, like Android, and will—in theory, anyway—work on any television. As Google says in its promotional materials, "Before long, anyone will be able to do anything with it." Not exactly comforting words to a worried parent.

So far, there are a limited number of applications for the Google TV platform, but they're enough to enable your child to stumble into some of the cybertraps discussed in this book. For instance, Google TV comes equipped with Twitter—admittedly, not the favorite social networking tool of teens and preteens, but one that could certainly be used to harass or bully classmates. More significantly, Google TV runs a version of Chrome, Google's Web browser, which gives users access to the full range of Web resources, including more common cyberbullying tools like Facebook and MySpace. Surveillance software is not among the applications currently available on Google TV, but it's possible to check the history of the Chrome browser to see how it's been used. Of course, it's also extremely easy to delete the browser history.

It's *What* They Do, Not *How* They Do It

From a parenting perspective, it doesn't really matter whether your child is a cyberbully by means of a laptop computer, a smart phone, an Xbox 360 console, or Google TV. The fundamental issue is not the technology itself, but the behaviors themselves that get kids into trouble.

As I've said throughout this book, parents have the primary responsibility for teaching their children cyberethics— it's an integral part of rearing children and, of course, the bulk of technology use takes place in the home. The main objective is to educate children from a young age about the appropriate boundaries and rules regarding the use of technology. Some of the lessons are not much more complicated than simple manners—don't take pictures of people without their permission, don't be mean to friends and classmates—but others are either more complicated or not fully understandable without more structured and long-running education and guidance. Some examples:

- Intellectual property theft
- Identity theft and computer hacking
- Online purchase of addictive substances
- Sexting and sextortion

The best approach is to work with your child to create a household "code of cyber conduct" that can evolve and adapt as your child matures or gets new electronic devices. This will give you an opportunity over a period of months and years to thoroughly educate your child about these more complex cybertraps. I also think it's appropriate for schools to include lessons about cyberethics and cybertraps in their curricula. Schools have a vested interest in minimizing the disruption that cybertraps can cause in the classroom, and many of the topics that naturally arise in a cyberethics discussion—intellectual honesty, cyberbullying, identity theft, hacking—are ones that directly affect schools and school personnel.

Regardless of the source of instruction, the bottom line is that children need to understand the specific cybertaps that exist, and the consequences of falling into them. For parents worried about how to keep pace with their children's innate grasp of technology and the rapid pace of change, it's important to remember that the basic lessons of human interaction don't change quickly. Courtesy, kindness, honesty, and respect are all persistent values. If those lessons are effectively taught, children will be able to apply them to whatever remarkable, convergent devices come along in their lifetimes.

Chapter Eighteen
DON'T Buy Your Child Technology Before They're Ready

Children are being exposed to—and gleefully adopting—technology at earlier and earlier ages. There are a number of reasons for this trend, all of which conspire against the efforts of parents to put on the brakes.

Many parents believe their children should learn to use technology as soon as possible, on the theory that a familiarity with technology is a necessary and important skill to have by the time children enter school. And of course manufacturers are naturally motivated to encourage this belief, and to expand the pool of potential consumers for their products—if six-year-olds like a mock laptop with pre-installed *Star Wars* educational software, why not something for four year olds, or even two year olds? Fisher-Price sells a toy it calls the Smart Screen Laptop, aimed at children between birth and twenty-four months, which features numbered buttons of different shapes and colors. When a button is pushed, the LED screen displays the corresponding number, and various sounds are played to entertain your infant. It's a remarkably short step from pushing the red "2" button to pressing an icon to launch the "Disney Fairies Fly" app (ages three and up) on an iPhone.

Meanwhile, electronic devices are getting easier and easier to use. After decades of relatively slow progress—from self-programmed devices to command-line prompts to early, balky versions of Microsoft Windows—computers, smartphones, and gaming consoles are now coalescing around the idea of the app. As Apple said when it released the iPad, "If you know how to use an iPhone, you already know how to use [an iPad]." Other smartphone and tablet manufacturers are following Apple's lead. Children, who seem wired to understand technology, find it laughably easy to use these new devices at ever-younger ages.

As kids get older and begin working their way through school, they confront the insidious effects of peer pressure. It used to be relatively rare for a seventh-grader to have a mobile phone. Now at least half of fifth-graders do—and they're not even the youngest mobile users. In the face of these trends, the parental "no" can take a beating. Don't let it. Resistance to requests for technology by those too young to understand the risks, or by those who have yet to demonstrate their maturity, is not being out of touch or unfair; it's parental responsibility.

With Technology Comes Responsibility

When considering whether to give your child a particular piece of technology, the most important question is fairly simple: is he or she mature enough to use it? Obviously, this is not a decision based solely on the number of candles on your child's last birthday cake. Instead it's an assessment of whether your child can understand what should and shouldn't be done with technology, as well as the consequences of crossing those boundaries.

Although it may be tempting (and in some cases appropriate) to issue a flat-out "no" to a specific technology request, it doesn't have to be an all-or-nothing proposition. There are compromises you can strike between your young child's understandable desire for the latest gadget and your concern for his or her well-being. For instance, using technology doesn't require having it available twenty-four hours a day. Take an inventory of the devices in your household and think about what time limitations make sense for you and your family. Discuss these restrictions with your child, make sure they are clearly understood, and stick to them.

Compromise may also mean resisting the incessant—and admittedly cool—feature-creep of modern technology. It's one thing for a third grader to have a digital camera, but is it necessary for a third-grader to have one with built-in Wi-Fi, easy connections to social networking sites, and e-mail? Similarly, you may decide there are perfectly good reasons for your sixth or seventh grader to have a mobile phone—safety and scheduling, for instance. But it's worth thinking seriously about whether your child really needs a phone capable of taking pictures, or a smartphone that can run multiple Web apps and comes equipped with built-in SMS (text messaging) and MMS (multimedia messaging) capability.

Avoiding feature-creep on more sophisticated devices like laptop and desktop computers is difficult, because even in their most basic configurations—an entry-level Compaq Presario, for instance, which retails for about $299—computers have all the capability necessary to trigger every cybertrap discussed in this book. Actually, the one thing missing is a built-in webcam for sexting, but that's easily fixed with the purchase of an inexpensive, peripheral hook-up.

Given the enormous educational benefits of computers and Internet access, it would be a disservice to our children to cut them off from computer usage altogether, and few parents would want to go that far. But there's absolutely no question that—for the moment, at least—computers pose the greatest risk for cyber trouble. Since it's virtually inevitable that your child will begin using a computer at a young age—particularly if he or she has older siblings—your efforts to educate your children should begin very early, meaning as soon as they can communicate.

Brace Yourself for Difficult Conversations

When I was a junior-high student in the mid-1970s, there was much gossip among my classmates about "the talk"—the halting, invariably uncomfortable conversation with our parents about the details of human reproduction. This was supplemented by an equally uncomfortable short film shown in our sixth-grade classroom, a half hour of gender-specific line drawings that made little or no sense at the time. Things got more detailed (and only marginally less uncomfortable) in our ninth-grade health class, but by senior year, a dozen or so of my classmates—including the prom king and queen—had become parents, so somehow the general concept of how sex works had gotten through, if not the lessons about its attendant consequences.

All of that, of course, was well before cable television, video-on-demand, the Web, and the explosion of the online adult industry. Most parents who sit their children down for today's version of "the talk" are dismayed to learn that their kids already have more than a passing knowledge of the birds and the bees. Technology is taking what used to be considered "adult" knowledge and making it available to younger and younger children. It will be some time before we can fully measure the impact of that change.

While child exposure to sexually explicit content is an important issue, it's primarily a moral and developmental issue, rather than a legal one. The greater concern, from a legal perspective, is how technology gives children the ability to commit crimes far beyond their years. In this day and age, the subject of "the talk" should no longer be limited to sex and its attendant risks. It also needs to include a discussion of boundaries that we've long taken for granted—privacy, decency, nudity, honesty, and other moral imperatives—which, thanks to technological advances, can easily be dismissed by children.

Few adults will look forward to having a frank discussion about nudity and the dangers of sexting with their ten-year-old son or daughter (their kids won't exactly be thrilled about it, either). But if your child is carrying around a smart phone with a camera that can be used to send a photo to dozens or hundreds of other kids, it's a conversation that needs to happen.

Chapter Nineteen
DO Require Full Friend Status on All Social Networks

Supervision Is Not Stalking

Even if our kids don't fully appreciate the consequences of the cybertraps discussed in this book, they're at least aware of the terminology. On more than one occasion, I've made a comment about something I saw on one of my children's Facebook walls that I found questionable, only to hear my child ask, "Why are you cyberstalking me?"

The answer is straightforward: I'm not cyberstalking you, I'm cyber-*supervising* you, which is a different thing altogether. The purpose of that supervision isn't to annoy our children or make them paranoid, but instead to offer them guidance and help them understand the consequences of what they say and do online. One of the biggest problems with the Internet—and one that's certainly not limited to kids—is that it has a tendency to distance us from the emotional reactions of the people we communicate with. When we're online, it's harder to tell if we've hurt or insulted someone by speaking carelessly. We can't see their facial expressions, and they can't see ours, nor is it always easy to convey tone—jokey or sarcastic, for instance—with keystrokes. At the same time, it's easier to say cruel things, or even get drawn into a so-called "flame war," exchanging incendiary remarks, when we don't have to look at our adversaries face to face.

We need to teach our kids that, unlike in Las Vegas, what someone says on the Internet does not always stay on the Internet. Children must come to terms with the fact that no matter how hard they try to be protect their online privacy, school administrators, coaches, college admissions officers, job recruiters, and employers will be looking at their online comments and activity and making judgments about what they see. Supervising what our children do online is part of a larger responsibility we have to educate them about contemporary privacy issues, and make them aware that the information they put online may travel much farther than intended.

You're the One Friend They Can Trust

A quarter-century ago, computer guru Stewart Brand famously said, "Information wants to be free." His point was that the emerging Internet would reduce the cost of sending information from point A to point B, but the communication revolution that's taken place in the intervening years has also given new meaning to his pronouncement; namely, that information has an organic quality and wants to escape from whatever device is first used to capture or create it. There are literally hundreds of teens (not to mention Hollywood celebrities) who have been shocked to discover that their supposedly private nude photos or sex tapes have wound up on the Web.

The implications of sending a message from a smartphone to another person, or even to a handful of friends, are relatively simple for kids to understand. What can be much more difficult is getting them to fully comprehend the exposure and embarrassment that can result from the one-to-many nature of online message boards, social networking sites, and the Web as a whole. The average teen, of course, will scoff, saying he or she perfectly understands that a Facebook post can be read by any other Facebook friend who has the correct privileges. Privacy settings, the teen will say,

are all that's needed to keep a bit of gossip or an embarrassing party photo from spreading across the globe. But it's not
that simple.

It's true that social networking sites like Facebook and
MySpace have vastly improved their handling of privacy issues,
and users have far more control than they used to over who
has access to material they post online. The essence of privacy,
after all, *is* the ability to control what information you share,
and with whom. Unfortunately, even with increased privacy
controls, social networking sites really only create the illusion
of online privacy. The reality is that it takes only one indiscreet
friend to copy something to his or her Facebook page, or to an
external website, or to a hard drive, for all your child's control over that content to effectively evaporate. And many social
networking sites—Facebook in particular—are pretty aggressive about using the information posted on their sites for their
own purposes, including product and site development, internal advertising, and, in some cases, sale to third parties.

There's a useful analogy you can employ to help your child
understand the potential spread of information online. With
proper privacy controls in place, posting photos to a Facebook
page is roughly the same as printing those photos and putting them on the family refrigerator. In theory, the only people
who will see them are those in the extended family circle—
parents, siblings, other relatives, friends, and friends' parents.
Your child may have the reasonable expectation that the photos won't spread beyond that circle. But they certainly could:
It would take only one friend using a mobile-phone camera to
snap a few quick copies and share those photos with others,
maybe even post them to a blog or his or her own Facebook
profile.

Suddenly, instead of being safely confined to your family
refrigerator, the embarrassing photos are available to a much
larger audience—anyone who walks by a school bulletin board,

for instance, or reads the local paper, or drives past the billboards at the edge of town. The same thing can be done with virtually any content your child posts online, which is one of the reasons the Internet is often referred to as "the global copy machine."

The goal is not to make our children reclusive or unduly reticent about sharing information with their friends. Instead, the objective is to make them stop and think, even briefly, about the consequences of what they post to the Web, and how they might feel if a wider audience were to see it.

Access, Passwords, and Full Friend Status

All this is moot, of course, if you can't actually see what your child is posting online. There are a couple ways to keep up with your child's online activity: collaboration and interaction, or electronic surveillance. In an ideal world, the first approach would work, but if your child is intrinsically resistant to it, or if you have reasonable concerns that you're not getting all the information you need, then the use of surveillance software may be the only reasonable solution.

In order to effectively supervise your child's online activity without electronic surveillance, you'll need physical access to whatever devices he or she uses, any passwords for those devices and for websites your child frequents, and "full friend status" on social networks. The details of full friend status may differ from site to site, but the basic concept is simple: assuming you have your own account on the site in question, make sure your child provides you with the fullest access possible to his or her information. Facebook, for instance, has introduced multiple categories of individuals with whom information is shared: Friends, Friends of Friends, and Everyone. Regardless of how your child might actually characterize you, you should be in the first group, which enables you to see everything your child posts and everything that is posted on his or her online "wall," the bulletin board that Facebook creates for each user. Most other social networks have a similar setup, so the same general principles apply.

If you make those conditions household rules—rules which must be met for your child to keep the privilege of using electronic devices—the resistance, at least at first, should be minimal. As children enter their teen years, they're more likely to complain about supervision, or to protest the idea of giving you unfettered access to their online activity. The problem, of course, is that the period of greatest resistance—middle school through high school—coincides with the period in which our children are at the greatest risk of falling into one or more cybertraps. You may be reasonably confident your child, by his or her middle school years, has absorbed the necessary cyberethics lessons to have earned more privacy, but it's inadvisable to abandon supervision altogether during that crucial developmental period.

One ongoing challenge in supervising your child's online activity is simply keeping track of all of the websites and social-networking services he or she joins, particularly as your child gets older. Kids are notoriously fickle, and it will take some work to maintain an up-to-date and accurate list, even with a fully cooperative child (just stop and think for a minute about the number of sites on which *you've* created a user ID and password). Many of the available software packages can relieve you (and your child) of the effort needed to keep track of all those sites by simply recording them for you. Moreover, since many of the packages have keystroke loggers built in, it's possible to compile a list of user IDs and passwords. Surveillance software may be the only solution if your child is uncooperative about passwords, or the details of his or her online activity. The farther your child gets from cooperation and effective partnership in his or her education about online behavior, the more it's worth considering whether the use of certain electronic devices should be curtailed or even eliminated.

Ultimately, the level of surveillance you choose will depend on the number of productive conversations you've had with your child about cyberethics, privacy, boundaries, and other aspects of safe online behavior. Yes, many of those conversations can be uncomfortable or awkward, particularly when children are still young. But children grow up fast online (or they think they do, anyway), and it's far easier to have those challenging conversations when a child is young than to wait until he or she is older.

Chapter Twenty
DON'T Allow Any Peer-to-Peer Software

Useless Until Proven Otherwise

I realize that many of the suggestions in this book have come with various options and caveats: if your child is young, you should consider this, but if your child is older, perhaps *this;* you can supervise your child this way, but you might also consider *these* options. Parenting is not an exact science: every child, and every household, is unique, so there are few hard and fast rules. However, this next rule is about as close to hard and fast as they come: do *not* permit the installation of peer-to-peer file-sharing software on any computer in your house. While there may be a handful of legal uses for such software, its primary purpose is to facilitate copyright infringement (a crime), and it's all too easy to either accidentally or intentionally download child pornography while using it (a much bigger crime).

There's absolutely no question that the most popular use of peer-to-peer software is to download pirated content. A 2008 study by the University of Hertfordshire in England found that the average iPod user between the ages of fourteen and twenty-four had 842 stolen songs on his or her device, roughly 48 percent of the average iPod library. When researchers limited the age group to those between fourteen and seventeen, the percentage of stolen music rose to 61 percent.

The statistics on illegal movie downloads are no less startling. *The Hollywood Reporter,* a movie-industry trade journal, estimated that James Cameron's hit *Avatar* had been illegally downloaded sixteen million times, making it the most pirated movie of 2010. All told, the movie industry estimates that the American economy loses $20 billion per year to video piracy, which is one of the reasons Hollywood has enlisted the help of U.S. Immigrations and Custom Enforcement (ICE) to crack down on the problem. ICE is currently running a program called Operation In Our Sites to seize Internet domains that host (or in some cases merely link to) pirated content. So far, ICE has seized more than eighty domain names and filed criminal complaints against a dozen domain operators for their roles in promoting theft.

Unless your child is actually operating a piracy site (which is unlikely), there's little chance of ICE showing up at your doorstep, since their focus is almost entirely on websites rather than individual users. However, the same is *not* true for copyright owners, who are more than happy to seek damages from individuals who pirate their copyrighted songs or movies. As I discussed in chapter 5, peer-to-peer software only works if the software knows the IP address of the individual downloading content. That makes it a relatively simple matter for Internet Service Providers to track who's downloading what and to provide that information to copyright owners (or, in the case of contraband, to law enforcement). Consider the warning posted on the website of Grokster, a peer-to-peer service based in the West Indies that was shut down in 2005 after the United States Supreme Court unanimously ruled it could be sued for copyright infringement:

> The United States Supreme Court unanimously confirmed that using this service to trade copyrighted material is illegal. Copying copyrighted motion picture and music files using unauthorized peer-to-

peer services is illegal and is prosecuted by copyright owners. There are legal services for downloading music and movies. This service is not one of them.

YOUR IP ADDRESS IS XXXX.XXXX.XXXX.XXXX AND HAS BEEN LOGGED.

Don't think you can't get caught. You are not anonymous.

Give Your Child the Chance to Convince You

There are some legitimate uses for peer-to-peer software. For instance, you can use it to download movies or songs that have entered the public domain. Many bands offer songs or music videos on P2P networks as a way of building an audience. Similarly, P2P networks are often used by game manufacturers to offer demos, and by open-source-software communities as a low-cost method for distributing free programs, patches, and software updates. There are even a handful of book publishers who distribute sample chapters or e-books via P2P networks.

If your child wants to install P2P software—or, more likely, wants to keep it after you've discovered it on the computer—ask him or her to explain in detail the non-criminal uses he or she is interested in. It's a great opportunity to discuss the potential cybertraps, from intellectual property theft to child pornography, and your child's answers may give you insight into how he or she is using the computer. It's also a chance to help your child understand the difference between content that seems free (because it's so easy to download and there seem to be no negative consequences) and content that actually is free.

This is, by the way, one of the areas in which a positive parental example can make a big difference. Kids have highly sensitive hypocrisy sensors, and their receptiveness to your warnings about copyright theft will be substantially diminished if your pirated copy of *The Essential Michael Jackson* is blasting from your computer speakers. And while it may be obvious, it's worth underscoring that all the cybertraps that can snare your child can also snare you.

Monitor It Closely

If your child succeeds in persuading you that he or she has a good reason for keeping P2P software on the computer, make it clear that your permission comes with a price: a much higher level of supervision. Given the real-time surveillance for contraband (i.e., child pornography) law enforcement can and does conduct on P2P networks, there's simply no margin for error. Even a single accidental download of child pornography can result in your home IP address being logged in a law-enforcement database, and it's far too easy to either intentionally or carelessly download enough contraband to trigger more intensive scrutiny.

Your first step will be taking the time to understand the software your child is using to access P2P networks. You can begin by asking your child to show you that software and explain how it works. But you'll want to supplement your child's explanation by Googling the name of the software and visiting its website. For instance, if your child is using FrostWire, a visit to that program's site will provide you with screen shots, detailed instructions on how to install and use the program, and a Frequently Asked Questions section about various aspects of the program (significantly, the only mention of the word "copyright" in the FrostWire FAQ is the link to the software company's own copyright statement).

It's also helpful to visit Google News or a similar news-search site and type in the name of the software or website you're researching. You'll quickly see whether a particular software suite or site is associated with any of the cybertraps discussed in this book. Plus, providing your child with specific examples of these potential problems is a good way to get his or her attention. For instance, in the last fourteen months, there have been just two publicized child pornography prosecutions featuring FrostWire, compared to the hundreds involving LimeWire. Now that LimeWire has been shut down, however, you can expect FrostWire to become a more common element in those reports.

Once you've familiarized yourself with the functions of the P2P software your child is using, the next step is to periodically check on the nature of the files he or she is downloading. Again, you can start by asking your child directly, but ideally you'll have learned enough about the software (and how to conduct basic searches for computer files) to verify the information your child gives you. It's important to remember—law enforcement certainly will—that there are myriad places on a computer where downloaded files can be stored; how aggressively you'll need to search for those other locations will depend on your assessment of your child's online activity. Since image files (typically stored with the extension *.jpg*) or video files (common extensions include .mpg, .avi, or .mov) can be stored anywhere on a computer, you'll want to either use the built-in search capabilities of the computer (both Windows and Mac OS can search entire hard drives), or download software specifically designed to index and search hard drives. The other thing to keep in mind, of course, is that computer files can easily be stored and accessed using an external hard drive or thumb drive, so keep those in mind as well.

The simplest and most thorough method for monitoring P2P activity is to use surveillance software. Most monitoring software suites can capture the searches your child uses within the P2P software, something the P2P software itself doesn't typically save from one session to the next.

You'll also be able to see the names and types of files your child downloads. Given the amount of information you can glean from surveillance software, it's tempting to let it do all the monitoring for you. But it works best when it's an open supplement to both your own interaction with your child, as well as your physical supervision of the electronic equipment in your household.

Chapter Twenty-One
DO Network with Friends, Teachers, and School District

It Takes a Village

It's reasonable for parents to feel overwhelmed by the task of supervising their children's electronic activities. The rapid pace of change, the inscrutable terminology and instructions, and the occasionally annoying facility kids demonstrate with new technology can all combine to make parents feel frustrated, impatient, and even a bit (dare I say it?) antiquated. It's not that adults don't use technology, but there's a subtle difference between using it and mind-melding with it, which is what most kids seem to do.

The problem is that children can't simply be left to their own devices. Too much of what takes place online has real-world consequences, which the majority of children are too young and too inexperienced to fully appreciate. Of course, it was certainly possible for children to make foolish and life-altering decisions prior to the rise of the Internet, but in our wired (and wireless) world, trouble happens much more quickly and easily.

As I've said throughout this book, parents have the primary responsibility for educating their children about potential cybertraps. But that responsibility can and should be shared—with other parents, with teachers and school administrators, and even with local law enforcement. The goal is to provide a unified front to children. Through the coordinated effort of the entire community we need to show our kids what behavior is acceptable and unacceptable, and help them to make the transition to adulthood free from the consequences of foolish or malicious online behavior.

Networking with Other Parents and Friends

The easiest place to start building your network of allies is with other parents in your neighborhood, or the parents of your child's classmates. Talk to these parents about technology issues in general and, in particular, the ways your children communicate with one another electronically. Find out what measures, if any, other parents have taken to protect their children, and what they've learned about the programs and devices their children have used. Ask about problems they've experienced and how they've dealt with them. Exchange phone numbers and e-mail addresses so you can keep in touch easily, make it clear that you welcome their help in supervising your own child's electronic activity, and ensure them you'll gladly return the favor.

A great place to do this type of networking is through the parent-teacher association (PTA) in your child's school. Once or twice a year, organize a technology-update evening to educate parents about the various cybertraps that can snare their children, new ways kids can get in trouble (a price of technological progress), and what can be done to protect kids in the community and in the school. If there's no one in the PTA who feels comfortable leading such a discussion, contact your local police department and ask them to send a representative to meet with you. They'll be more than happy to do so.

A little over a year ago, I collaborated with my local state's attorney, T.J. Donovan, and a member of the Vermont Internet Crimes Against Children task force, detective Kris Carlson, to put on that type of information session for Burlington parents and educators. The meeting was particularly well attended by school social workers, who said they were seeing a variety of behavioral issues in schools relating to online activity. Much of their concern centered on the taking and distributing of inappropriate photos, a point that Carlson deftly illustrated by surreptitiously taking a photo of me during the lecture and, within seconds, emailing it to my computer. For the social workers and other administration officials, it was a sobering look at how quickly technology is changing and the issues that it is raising both in and out of school.

Networking with Teachers

The next layer of protection for your child comes from his or her teachers. Because they see your child on a near-daily basis, they're in a great position to observe the kinds of behavioral changes and classmate interactions that could signal a potential problem. When you meet with your child's teachers, tell them you welcome their help in keeping an eye on your child. Make it clear that you don't view them as substitutes for your own supervision, but as partners who can play valuable roles in protecting your child from cybertraps.

Remember, however, that teachers interact with dozens of children every day, and it's not easy to keep track of each child's shifting moods, let alone the fluid social interactions that characterize preteen and teen relationships. As a school-board member, and someone who has had a number of public school teachers in my own family, I'm very much aware of the demands that have been placed on teachers over the years— many of which go far beyond their primary goal of educating our children. Monitoring the emotional health and online activity of students should not be a requirement of a teacher's job; on the other hand, it's reasonable to expect that if they observe cyberbullying, sexting, or other inappropriate technology use, they'll relay that information to school administrators or directly to you. In fact, the Burlington School District has adopted a policy that requires any school employee who witnesses or hears of harassment to "take reasonable action to stop the conduct and prevent its recurrence," including reporting it to a designated school official.

Of course, the more information teachers have about potential cybertraps, the more effective partners they can be in protecting children in their classrooms. If you have information about a particular cybertrap that concerns you, share that information with the teacher (or, better yet, hand over a copy of this book). Invite teachers to PTA training sessions where they can hear the concerns of parents in the school community and learn more about possible cybertraps—an added benefit to this approach is that many teachers are parents themselves. Much of the information that's available from student resource officers or local law enforcement can be easily incorporated into lesson plans for the benefit of all students.

Networking with School Administrators, Law Enforcement

Given the disruption that cybertraps can cause in schools and the broader community, it's not surprising that many school administrators and police departments have begun to take them seriously. Neither school administrators nor law-enforcement officers are typically in a position to monitor your child's online activities directly, but they are in a position to help educate your child about the risks of cybertraps and create a safe educational environment.

Find out if your school board or district has adopted specific policies regulating the use and misuse of electronic devices. For instance, the Burlington School Board has adopted two relevant policies, one covering acceptable in-school use of electronics, the other handling student harassment (both policies are discussed in greater detail in chapter 22). Your school district's policies should be available on its website, but if not, ask the principal to obtain copies for you, or contact the district's central office.

If you discover that your district doesn't have specific policies, consider launching a campaign to create them. The best place to start is with a local school-board member. If the board is reluctant to take up the issue, mobilize support by talking to neighbors and other parents in your school community. Take advantage of public comment opportunities at school board meetings to advocate for these policies. If necessary, run for the board yourself. There are far worse issues on which to base a campaign.

Depending on how your local school board functions, it may either instruct a subcommittee to draft specific policies or ask the board's attorney to do so. (A variety of samples of these policies can be found online at www.cybertrapsfortheyoung. com). New policies typically have to go through two or three readings before they can be adopted and implemented by the district administration.

Part of any district policy should be periodic training for the entire school community—administrators, teachers, students, and parents—about possible cybertraps and how to protect against them. There are a variety of ways in which this information can be disseminated: as part of the curriculum, during student assemblies, through in-service training, or as part of parent information nights or newsletters to the community. The important thing is that everyone understand the potential risks and the role the entire community needs to play in protecting children from unintended consequences.

When it comes to helping children understand those consequences, it's tough to beat the impact of a police officer in uniform. If your child's school or the district as a whole has a student resource officer, he or she has almost certainly received training on the types of cybertraps discussed in this book, and can effectively lecture about them to students and staff members. If not, it's likely that your local police department—or the state police—has someone who can discuss the topic with teachers, parents, and students. There are few things more sobering than listening to an officer describe arrests he or she has made, or computers he or she has had to seize, because a child did something illegal online.

Chapter Twenty-Two
DO Know the Policies and Laws in Your Area

Educate Yourself from the Inside Out

Much like the networking discussed in the previous chapter, educating yourself about the policies and laws that apply to your child is a process that starts in the home and works outward.

Start by reviewing your own household rules about the use of technology. Have you had an age-appropriate conversation with your child about the possible cybertraps he or she might encounter? What are the consequences if your child misuses electronic devices? What time limits, if any, have you placed on the use of particular devices? Are these rules written down? How often are they reviewed?

It cannot be stressed enough that the primary responsibility for educating your children about cybertraps—and protecting them from their potentially serious consequences—lies with you, the parent(s). Parents are the ones who know their children best, and are in the best position to provide them with the necessary guidance and supervision.

Moreover, you can tailor your household rules in such a way as to provide a balance between your child's maturity and the risks he or she might be facing, and you can adjust those rules as your child matures. Neither schools nor governments—the other main promulgators of policies and laws in this area—have the luxury of focusing exclusively on *your* child's best interests and capabilities.

School Policies and Procedures

As I discussed in the opening chapters of this book, a significant percentage of children in the United States have experience with personal computers before they even enter kindergarten, so hopefully you've already had some conversations with your child about his or her online behavior. But the real conversations about cyberethics will need to start once your child enters formal schooling, either as a kindergartener or first grader.

Computers have become an integral part of the curriculum at all levels of schooling. If your school district is one of the many in the U.S. with an "acceptable use" policy for electronic media, your child will be subject to it the minute he or she walks through the door. The same is true for any policies that prohibit specific types of electronic misbehavior.

Most school districts have some type of policy regarding the acceptable use of electronic resources and, in particular, the types of content that can be accessed on the Internet. Under the terms of the Children's Internet Protection Act (CIPA), adopted by Congress in 2000, any school (or library) that receives so-called "e-rate discounts" for phone service and Internet access must install a "technology protection measure" on any device capable of connecting to the Web. Schools are required to block access to various types of material (obscenity, child pornography, and material "harmful to minors") on those devices and to adopt an Internet safety policy.

The Burlington School District's policies offer a couple useful examples. For instance, the district's acceptable-use policy is designed to lay out rules for using electronic resources owned by the school. As required by CIPA, the policy provides that all students who use school-owned electronic resources "will receive instruction regarding the safe, ethical, legal, and responsible use" of those resources, as well as the Internet more generally. Under the Burlington policy, students cannot access materials for any purpose the district deems potentially harmful, inappropriate, illegal, or non-educational. The policy also aims to educate students about the appropriate use of the various personal devices they might carry around with them.

While in most cases the school doesn't have the authority to regulate what kids do off-campus, the district's anti-harassment policy makes clear that it *can* regulate off-campus conduct when it affects the learning environment within a school. The classic example is a sexting photograph taken in a child's bedroom that leads to on-campus harassment of the child, or distracts other students from schoolwork. Unless the photo was taken with school equipment, the district cannot punish a student for taking the photo, but it can discipline a student for being a harasser in school, disrupting the day by showing photos on a mobile phone, or distributing a sexting photo across the district's computer system.

As your child enters school, you should review the policies that have been adopted in your own district. You can typically find links to these policies on the district's website. If the policies aren't readily available, contact the district's central office for copies (they're public records) and suggest they be posted online. Make sure you refer to these policies when educating your child about acceptable and unacceptable online behavior.

If it turns out that your child's school district does not have an acceptable-use policy, that's definitely an issue to raise with your local school-board representative or with the district administration. The goal is not to create additional cybertraps for your child, but instead to encourage the school district to educate itself and the entire school community about existing cybertraps and how best to avoid them. The most powerful part of any policy is not the prohibition, but the education designed to prevent violations.

State and Federal Laws

The reason education is so important, both at home and in the schools, is that state and federal statutes governing many of the cybertraps discussed in this book are not educational, but punitive. If children violate one or more laws because they haven't been properly educated, that's not of much concern to state or federal law-enforcement officers. Their focus, understandably, is on investigating and punishing violations of criminal law.

There are too many state and federal statutes governing cybertraps to include them all in this book without turning it into a legal treatise. However, to help make things easier for both parents and educators, I've posted a list of potentially relevant statutes on the book's website at www.cybertrapsfortheyoung.com. The statutes are organized by cybertrap and by jurisdiction (in each of the fifty states and in the United States as a whole), and I've included a brief summary of each statute's main provisions.

If you think your child may be charged with a cyber crime, or sued due to his or her online activity, you should consult with an attorney licensed to practice in the state in which your child resides. (If it's a federal crime, find an attorney who is admitted in the district where the charges were filed). The attorney will be able to provide you with the most up-to-date information on the relevant law and the potential consequences. Never rely solely on information that you find on the Internet; it is, at best, a starting place for further analysis and discussion with a trained professional.

Chapter Twenty-Three
DO Talk to Every Child About Possible Cybertraps

In "Ode on a Distant Prospect of Eton College," Thomas Gray's melancholy rumination on lost childhood, he closes with a familiar phrase: "[W]here ignorance is bliss, / 'Tis folly to be wise." His point is that for children, life's hardships will rear their ugly heads all too soon, and it doesn't profit them to know too soon what life holds in store:

Alas! regardless of their doom,
The little victims play;
No sense have they of ills to come,
Nor care beyond to-day:
Yet see how all around 'em wait
The ministers of human fate
And black Misfortune's baleful train!

Gray's exhortation to shield children from the realities of the adult world is practical in certain arenas, but not in others. Remember: Gray wrote his poem in 1747, long before the invention of the computer, the mobile phone, or the Internet. The simple truth of the matter is that if we are going to give our children adult equipment, we should be prepared to have serious conversations with them about the adult consequences that can result from its misuse.

There are many reasons parents don't want to have those conversations: discomfiture or embarrassment about the sexual elements of many cybertraps, confusion or ignorance regarding the technology involved, nervousness about suggesting behaviors that haven't yet occurred to their children, or simple regret at the idea of robbing their children of the innocence that makes childhood so precious. The reluctance is entirely understandable—no one wants to hasten their child's introduction to "black Misfortune's baleful train." But parents have a unique opportunity to help ease their children's passages into adulthood, and to help them avoid the cybertraps that can painfully accelerate that journey.

It All Comes Down to Communication—Who's Saying What to Whom?

Each and every day, children are having conversations that influence their values and behavior. Many of these conversations are one sided, consisting of what children see on television or the Internet. Others are more interactive: with siblings, friends, and teachers. Others are entirely internal. In some cases, these conversations will mirror the values you hope to impart to your child, but often they are askance or in outright opposition to those values.

As director Woody Allen wisely observed, "Eighty percent of success is just showing up." If there are ethical or moral values you want your child to exhibit, the best way to promote them is to be a regular part of your child's conversations. There's no question that setting a good example is an important part of the mix, but an implicit message of cyberethical behavior is all too easily drowned out by the many loud and explicit messages competing for your child's attention. Don't be bashful about articulating your beliefs and expectations as clearly as possible.

Make Your Boundaries and Consequences Clear— Then Enforce Them!

Your child may not like it, but part of your conversation should be about setting appropriate limits on the use of electronic devices. One of the more useful ways to balance the scale— making sure the ethical messages your child gets from others don't overshadow your own—is to put a cap on the time he or she spends listening to competing voices, particularly those in mass media. Even with your best effort you won't be able to shut out those voices altogether, but it's reasonable to restrict your child's media consumption to a level somewhere below the national average of seven or more hours per day.

The next step is to be clear with your child about the consequences for breaking a rule, or falling into one of the cybertraps discussed in this book. The hope, obviously, is that your child won't jump straight to one of the more serious criminal cybertraps, which carry penalties beyond whatever you enforce at home. A more likely scenario is that your child will be caught saying something unkind about another child on Facebook, or downloading pirated songs. The question to consider beforehand is how best to respond to these lesser infractions so you can prevent them from turning into more serious ones. It can be challenging to figure out the right balance between the need to teach your child a lesson and his or her legitimate need to continue using electronic devices for homework, research, scheduling, and communication with you and his or her peers.

Whatever you do, don't give up. As most parents quickly learn, it's important to choose a consequence you can live with. If you take your daughter's computer away for a week, she may need to use yours in order to do her homework, which can be frustrating for both of you. If you declare that she's lost her mobile-phone privileges for a few days because of a 3:00 AM texting session with her girlfriends, you may have a harder time keeping track of where she is and what she's doing.

There are, unfortunately, no simple answers. Striking the proper balance between discipline and other important home values like safety, supervision, and convenience is complicated. The first and best step is prevention—engaging your child in ongoing, detailed conversations about your expectations. But in the event of inevitable breaches of those expectations, one approach may be to choose consequences that actually make your life easier. For instance, when it comes to relatively minor breaches, your child might be assigned additional chores—laundry, dishes, yard work, etc. More serious breaches might result in the loss of social events like movie outings, parties, or sleepovers. It's definitely a process of trial and error, and one that needs to undergo constant revision to reflect the changes in your child's maturity, sense of responsibility, social activities, and technology use. The more your child is a partner in setting boundaries and developing consequences that work for both of you, the less likely it is that he or she will be an adversary.

Early Warnings Are More Effective Than Cleanup

We tend to credit Ben Franklin with the words "an ounce of prevention is worth a pound of cure," but the saying actually dates back at least eight hundred years. A British judge and legal scholar named Henry De Bracton included the maxim in his 1240 treatise *De Legibus et consuetudinibus Angliae* (On the Laws and Customs of England). Nearly 1,300 years before that, the Roman poet Persius gave similar advice to his readers when he warned them to "meet the malady on its way." None of these writers foresaw the arrival of the Internet (although I suspect Franklin would have loved it), but they would no doubt agree that the old truism still applies.

The advice is good: it's far better (and cheaper) to educate your child about the possible cybertraps he or she might encounter than it is to resolve matters after the fact. Effective ed-

ucation begins in the home, but it requires the active cooperation of all the people in a child's life: family, friends, neighbors,
teachers, even school administrators and law enforcement.
While the Internet, like every human activity, does have serious traps that can snag an unprepared or unsupervised child,
there are numerous resources available to educate you about
the potential legal problems it presents, while giving you the
tools you need to instruct, guide, and protect your child. If
you've made it this far into the book, you're already well on
your way.

My sons, born in 1993 and 1995, are at the leading edge
of the post-Web generation. Like other children around the
country, they are true digital natives, cyber whiz kids at ease
with rapidly shifting technology and digital content delivery.
Children today are growing up in a world far different than the
ones in which their parents, or even older siblings, were reared.
The cybertraps discussed in this book are unquestionably serious and deserving of your and your child's attention. But they
are only a small slice of the total online experience.

The Web offers remarkable resources for communication,
education, entertainment, and personal development. My hope
is that this book will encourage you to talk at length with your
child about both the risks and rewards of this ever-changing
electronic world, so that each child can travel untrapped into a
productive and exciting adulthood.

Part Four: Resources & Tools

Appendix A
Electronic Monitoring Software

Many parents have viscerally negative reactions to the idea of conducting electronic monitoring of their children's online activity, and for good reason. It can create an atmosphere of distrust. It might serve as yet another flashpoint of disagreement with already-peevish teens and pre-teens. And surveillance, to many of us, seems downright un-American.

All of which is true. Yet despite my strong belief in the right to personal privacy, I think parents should give serious consideration to the judicious use of surveillance software.

It's worth remembering that kids grow up with many forms of surveillance: we watch toddlers to make sure they don't run into the street or put something dangerous in their mouths; we watch young children and preteens around the kitchen, as they learn to use the stove and cut vegetables; and we watch our children's every move as they learn to drive. In each of these instances, there are basic lessons that need to be absorbed and, until we're confident that our children have learned them, we watch (and worry).

Kids these days, more often than not, are teaching themselves to use electronic devices, with little or no adult supervision. As a result, they're not necessarily getting the moral and legal guidance they need to keep themselves safe on the information superhighway. Many of the cybertraps discussed in this book are unfamiliar to teens, or else difficult for them to fully understand. Of course it's impossible to stand over our children's shoulders every minute they're online, but I think it makes sense to employ the help of available technology to help protect them from harm.

Filtering or Monitoring?

In the early days of the Internet, when parents first became aware of the objectionable content available online, a number of companies introduced software that attempted to block sites containing some of the usual suspects: sex, drugs, alcohol, violence. These relatively unsophisticated programs used one of two approaches: a constantly updated "black list" containing hundreds of thousands of forbidden web addresses, or lists of forbidden keywords that, if found on a site, would restrict a user's access. Neither option was perfect, or even close to perfect. Keyword blockers, for instance, often cut off access to perfectly benign sites, like web pages containing recipes for "chicken breasts."

Since those early days of the Web, filtering/blocking software has gotten considerably more sophisticated. Most packages now use a combination of techniques to identify objectionable material; some are even sensitive enough to permit a site in the morning and, if the content changes, block it in the afternoon.

Though the technology has drastically improved, my preference is for software designed to monitor and record *all* Internet activity, rather than one that tries to block specific sites. In my view, anything less than a comprehensive record opens the possibility that your child could find a way around the filtering algorithms, and, in fact, your child might take finding such loopholes as a personal challenge. Monitoring software will give you the full picture of your child's online activity, and will leave it up to you to decide how to respond to activity you might find troublesome.

1. Hidden or Open?

One of the most popular features offered by monitoring packages is the ability to conduct stealth supervision of your child's online activity. The appeal of hidden monitoring is understandable: many parents feel that it's the best way to learn what their children are really up to online. But whether it's the government, businesses, or parents, I think hidden monitoring is ultimately unproductive and intrinsically demeaning.

For starters, the effectiveness of hidden monitoring is hampered by the fact that it remains hidden only until you learn of some activity you need to discuss with your child. The inevitable question, "How did you know *that*?" will effectively blow the software's cover, and you'll find yourself dealing with a highly upset and distrustful child.

A much better approach, I think, is to tell your child upfront that you'll be installing the software. The very knowledge of its existence may curb problematic Internet behavior, and it can also present an opportunity for communication. You and your child can collaborate on selecting an appropriate package and installing it together. You can discuss the reasons you feel monitoring is necessary, and you can begin to gauge his or her understanding of various cybertraps. The two of you can talk about the extent of your monitoring, and you can lay down some conditions for when you might consider easing or even eliminating your surveillance.

2. Judging the Effectiveness of Monitoring Software

Listed below are some of the more common features available as part of monitoring software packages, along with a brief description of how each one works and how useful it is in keeping track of your child's online activity. Keep in mind that the manufacturers are constantly updating their software, so it's a good idea to conduct a little additional research before purchasing a particular package.

There are a number of websites that can provide you with in-depth reviews of the available options; the ones I find particularly useful are TopTenReviews (http://www.toptenreviews.com/), Monitoring Software Reviews (http://www.monitoringsoftwarereviews.org/), CNET (http://download.cnet.com/windows/monitoring-software/), and Keylogger.org (http://www.keylogger.org/).

Thoroughness of Monitoring

When you're reviewing monitoring software, check to see what types of sites the software monitors and/or filters. Are there types of online activity that the software does *not* monitor, such as social networking, video, peer-to-peer sharing, or gaming? How effectively does the software handle non-Web traffic, like e-mail or chat?

Keystroke Logging

The basic idea behind keystroke logging is that it allows you to view everything your child does on a computer, regardless of what tools or transmission protocols he or she uses. Many packages have an option to record every keystroke your child enters on the computer. A good question to ask: does the software include tools for organizing that keystroke information in a relatively readable form? Otherwise it might look like nothing more than a long string of indecipherable gibberish.

Storage and Transmission of Activity Reports

Even the best monitoring software won't be much help if you can't see what's been recorded. Some packages are better than others at organizing the information and presenting it in a readable form. Often, you'll have several options for viewing reports: logging on to the monitored computer, receiving e-mails directly to your inbox, or accessing reports through the company's website. I generally find e-mail transmission the most convenient option, but you can experiment with different approaches and see what works best for you.

Screen Shots

Most monitoring packages now have the capability to take periodic screen shots of your child's computer activity and store those photos as part of a report. That can be particularly helpful, as it will give you a visual of what your kids are looking at. Keep in mind, however, that screen shots will take up a lot more room in the activity log than other data, which may become an issue in terms of storage space or the time required to transmit activity logs from your child's computer to either your e-mail or the manufacturer's website.

Data Filtering and Mining

Depending on how much time your child spends on the Internet, the activity report generated by the monitoring software can be quite lengthy. When you're reviewing the various packages, be sure to look at the report formats. For instance, is Internet activity broken into categories (a section for chat, another for e-mail, another for Web surfing, and so on)? What tools are available for searching within the activity report? The more resources the program offers, the easier it will be for you to home in on particular areas of concern.

Site Blocking

Despite the inherent challenges, most monitoring software packages do offer the option of blocking (or at least trying to block) objectionable content in various categories. You may find that feature useful to help screen your children from the most obvious types of disturbing or upsetting content. But be aware of the limitations, and expect that the filtering/blocking feature will grow less useful as your child gets older and more computer savvy. It's also worth considering whether the presence of blocking or filtering software will encourage your child to go somewhere else for his or her curiosity-driven explorations, which in the end will only make it more difficult to keep track of what you child is doing.

Appendix B
Glossary of Computer, IM, and TXT Terms

Texting, Internet messaging, and chat rooms sometimes feel as if they were specifically designed with children in mind: the instantaneous transmission, the short message length, and the rapid information exchange are characteristics that generally appeal to the under-twenty crowd. In part because of the space limitations (particularly of text messaging) and in part because kids love having their own special language, these electronic communications are often filled with abbreviations and contractions that can be difficult to decipher.

Most of the text expressions discussed in this appendix fall into the category of "shorthand," i.e., a sequence of letters used to abbreviate a longer word or phrase. For instance, TTFN is shorthand for "ta-ta for now." Sometimes these expressions are actually pronounced as new words. The old Army acronym FUBAR, for instance, is shorthand for "F*#%ed up beyond all repair" and is pronounced "foobar."

This kind of shortening has a long history in computer circles. Back in the day when every bit was precious, programmers used abbreviations as a way of saving keystrokes and transmission time. Over the years, numerous Internet-related initiallisms have crept into mainstream use: FYI ("for your information"), FAQ ("frequently asked questions"), BTW ("by the way"), BRB ("be right back"), LOL ("laugh[ing] out loud"), and of course, WTF ("what the @*%#").

The popularity of short-form communication among children (particularly texting) has caused a lot of concern among parents and educators, who worry that the practice is damaging the ability of teens to spell or punctuate properly, let alone form proper sentences. On that front, at least, there is some good news. According to a recent study by Coventry University, published in February 2011 in the Journal of Computer-Assisted Learning, there's no evidence that the literary skills of children are damaged by exposure to mobile devices and frequent texting.

In fact, researchers concluded that texting made a "significant contribution" to the spelling ability of the nine- and ten-year-olds who participated in the study. They speculated that the mental process of creating and decoding shorthand and "textisms" was similar to the process of understanding full-length words in literature. More broadly, researchers said, texting is a form of reading and writing, and contributes to the children's overall understanding.

Online Dictionaries of Abbreviations

The surge in texting by teens and pre-teens has raised another parental concern: that their children are using coded communication to conceal illegal or dangerous activity online. Various media organizations have stoked those fears by publishing lists of supposedly popular "acronyms" and stories of children tricked by Internet predators into disclosing personal information and private photos. In early 2009, the Fox News television station in Atlanta, Georgia earned considerable online derision, mostly from teens, for its list of "Top_50_Text_Acronyms_Parents_Should_Know" (not surprisingly, the page is no longer available on the station's website).

Part of the reason most kids are scornful of such lists is that, increasingly, teens don't actually bother abbreviating words in their texts, status updates, IMs, chats, or tweets. The teens I've spoken to about this issue say that using too much shorthand in a message is, to put it politely, "dorky." Moreover, they can text so quickly that it actually takes them longer to remember or construct an acronym or shorthand expression than to simply write words out at full-length.

YMMV ("your mileage may vary"), but if you regularly review your child's electronic communication, there's still a chance you'll stumble across some odd-looking strings of letters and symbols. Keep in mind, of course, that sometimes children are just messing with adult minds, and will create odd-looking words or acronyms to be amusing or even purposely scary. Regardless, there are some useful online resources to help you decode whatever cryptic messages you might find on your child's phone.

One particularly useful list, with a nice layout and clear explanations, is maintained by **Safe Internet Surfing** (http://www.safesurfingkids.com/chat_room_internet_acronyms.htm). It's thorough without being overwhelming, and offers some additional guides for interpreting unlisted shorthand words and emoticons (typographical abbreviations like :) [smile], ;) [wink], <3 [love], :([sad], and so on.).

The Computer, Tech, Web and Chat Term page of **Web-Friend.com** at (http://www.web-friend.com/help/lingo/) offers another nice collection of linguistic resources. In addition to a list of simple definitions of teen-oriented text and chat abbreviations, the site also provides information about a wide range of other specialized computer words, abbreviations, and symbols.

NetLingo.com is one of the oldest Internet dictionaries, with tens of thousands of definitions of computer and Web-related words. It bills itself as having "the largest list of text and chat acronyms" and that may well be right. Browsing through the list helps explain the mocking skepticism some teens have about the actual popularity of Internet shorthand. How intuitive is it, for instance, that "NC" stands for "nice crib," or that "NICE" supposedly stands for "nonsense in crappy existence"? Maybe the best summary is YYSSW— "yeah, yeah, sure, sure, whatever."

For an eye-opening and NSFW ("not safe for work") look at emerging slang, including many of the expressions discussed here, visit **UrbanDictionary.com**, a site that attempts to collect and define every new word and expression. Anyone can write his or her own definition and include it on the site, which means there are lots of obscure phrases you're unlikely to ever encounter in the real world, but there is a certain raw immediacy to the content.

The website **Teen Chat Decoder** takes a slightly different approach: rather than presenting a seemingly endless list of acronyms, the site provides a simple search box. Just type in the acronym you're trying to understand, and the site will suggest various possibilities. For instance, type in the characters "f2" and the search engine responds with F2F (face to face), f2p (free to play), and F2T (free to talk). It might be a little tedious to type in dozens of acronyms, but the search engine seems quite rapid and thorough.

Appendix C
A Brief Guide to Online Safety Resources

The vast majority of child-safety websites are designed to educate parents on how to protect their children from becoming victims of Internet predators. Taken together, they provide valuable lessons and warnings about the all-too-real harm that children can suffer from those who would take advantage of or abuse them.

The drawback to these sites is that they offer little or no information about the ways your child could intentionally or accidentally commit a cybercrime. The companion website for this book, cybertrapsfortheyoung.com, is specifically designed to address that need.

Among the features you'll find on the site:

- Headlines of articles illustrating the types of cybertraps that can snare children
- Links to relevant state and federal statutes related to the cybertraps discussed in the book
- Updates on proposed legislative changes meant to improve the application of existing laws to children
- Information on how to locate a criminal defense lawyer in your area links to other safety resources
- Links to reviews and comparisons of monitoring software

The purpose of the site is not to replicate the book you're holding in your hands, but to be a companion piece, providing detailed information that would have made the book itself too dense, as well as updates on happenings since the book went to press.

Below, I've listed several informative Internet safety sites you may want to peruse. While I expect that most of these sites will continue to exist for some time into the future, you can check the link list on cybertrapsfortheyoung.com for the most up-to-date safety resources. The information contained under each entry is quoted directly from the "About Us" section of each site.

1. **Chatdanger (http://www.chatdanger.com/)** "The Chatdanger website has been created to inform young people about the potential dangers and ways of keeping safe in interactive areas online, such as chatrooms, instant messenger, online games and e-mail, and also via mobile phones. The aim of the site is very much to inform and empower users of these services, so they can use these services safely, and not at all to discourage people from using these services."

2. **ConnectSafely (http://www.connectsafely.org/)** "Connect-Safely is for parents, teens, educators, advocates—everyone engaged in and interested in the impact of the social Web. The user-driven, all-media, multi-platform, fixed and mobile social Web is a big part of young people's lives, and this is the central space—linked to from social networks across the Web—for learning about safe, civil use of Web 2.0 together. Our forum is also designed to give teens and parents a voice in the public discussion about youth online safety begun back in the '90s. ConnectSafely also has all kinds of social-media safety tips for teens and parents, the latest youth-tech news, and many other resources."

3. GetNetWise (http://www.getnetwise.org/) "GetNetWise is a public service brought to you by Internet industry corporations and public interest organizations to help ensure that Internet users have safe, constructive, and educational or entertaining online experiences. The GetNetWise coalition wants Internet users to be just 'one click away' from the resources they need to make informed decisions about their and their family's use of the Internet. . .GetNetWise is a project of the Internet Education Foundation."

4. Microsoft Safety & Security Center (http://www.micro-soft.com/security/default.aspx) "This site offers a variety of information for families about online safety, including tips on creating strong passwords, avoiding cyberbullying, using social networking properly, and minimizing the risk of online predators, etc. The company offers a number of free downloads of security tools."

5. Wired Safety (http://www.wiredsafety.org/) "WiredSafety is a 501(c)(3) program and the largest online safety, education and help group in the world. We are a cyber-neighborhood watch, and operate worldwide in cyberspace through our more than nine thousand volunteers worldwide. (WiredSafety is run entirely by volunteers.)

"Our work falls into four major areas: help for online victims of cybercrime and harassment assisting law enforcement worldwide on preventing and investigating cybercrimes education providing information on all aspects of online safety, privacy and security.

"Together with our affiliate, www.wiredcops.org, specially-trained volunteers patrol the Internet looking for child pornography, child molesters and cyberstalkers. We also offer a wide variety of educational and help services to the Internet community at large. Other volunteers find and review family-friendly websites, filter software products and Internet services. . .

"WiredSafety is headed by Parry Aftab (also a volunteer), a mom, international cyberspace privacy and security lawyer and children's advocate. Aftab is the author of The Parent's Guide to Protecting Your Children in Cyberspace (McGraw-Hill), which has been adapted and translated around the world."

6. National Center for Missing and Exploited Children (http://www.missingkids.com/) "The National Center for Missing & Exploited Children (NCMEC) is a private, (501)(c) (3) nonprofit organization which was created in 1984. The mission of the organization is to serve as the nation's resource on the issues of missing and sexually exploited children. The organization provides information and resources to law enforcement, parents, children including child victims as well as other professionals."

7. SafeKids.com (http://www.safekids.com/) "SafeKids.com is one of the oldest and most enduring sites for Internet safety. It's creator, Larry Magid, is the author of the original 1994 brochure, 'Child Safety on the Information Highway,' and is also a technology journalist."

8. SafeTeens.com (http://www.safeteens.com/) "SafeTeens. com is a place for teens and their parents to learn safe, civil and responsible use of the Internet. It's operated by technology journalist Larry Magid who also operates SafeKids.com and is co-director of ConnectSafely.org."

9. Web Wise Kids (http://www.webwisekids.org/) "For more than ten years, [Web Wise Kids] has been providing unique and effective resources to equip young people to safely use and enjoy the latest technologies. Our programs prepare kids to be their own first lines of defense so they can confidently explore the best that the Internet has to offer. To date, more than eight million middle school and high school students from all fifty states have participated in our programs.

"Our programs assist youth to confidently manage issues like sexting, bullying, piracy, fraud, online romances, cyber stalking, and other online hazards. We take a 'hands-on, minds-on' approach to education by offering challenging and realistic digital games that have been specially designed to assist young people to evaluate their online activities and experiences, and take appropriate actions to stay safe online. Simulations that demonstrate the importance of digital citizenship and allow students the opportunity to witness the consequences of poor online choices are an integral component of all our programs."

Acknowledgments

"The profession of book writing makes horse racing seem like a solid, stable business."

—John Steinbeck

Parents are not the only ones flummoxed by the pace of change. It's difficult to imagine what Steinbeck would have thought of the book publishing business in a post-Internet world, as it grudgingly adapts to distribution methods almost unimaginable five years ago. It's a profoundly challenging time for writers, editors, and publishers, and the outcome is by no means certain for those of us who value long-form writing and the tactile sensation of books themselves. (One of the downsides to e-books is that they make all books feel the same. I can distinguish between my well-worn beach-bag copy of *The Hunt for Red October* and my treasured hardcover copy of *All The President's Men* with my eyes closed, but that day is fading.)

At such a time and in such an environment, it takes dedication to bring a new and challenging project to fruition. I would like to express my sincere appreciation to Jeff Link, my editor at NTI Upstream in Chicago, for seeing the potential and the need for this book. Throughout the writing process, Jeff offered insightful comments and constructive encouragement that helped make this a much better work. I'd also like to extend my thanks to Mike Ingram, the editor Jeff hired to work directly on my manuscript. Mike did a superb job not only with the mechanical aspects of the manuscript, but also with the overall flow and structure. He posed a number of thoughtful questions about the text itself, and pointed out places where additional material would make the book more thorough and more understandable.

I think that serendipity plays a role in most publications, and this book is no exception. At the 2010 National School Boards Association (NSBA) annual conference in Chicago, I attended a good presentation on cyberbullying put on by two Chicago attorneys, Eric Sacks and Sarah Terman, both of whom are affiliated with the Anti-Defamation League. When I was circulating the proposal for Cybertraps for the Young and looking for a publisher, I sent the file to both Eric and Sarah in the hopes they might know someone interested. Sarah wrote back and said she knew someone at NTI Upstream—fortunately, it turned out that her husband, Jeff, was in fact interested. So my thanks to Sarah for a great introduction.

It is appropriate that the origin of this book stems in part from my trip to the NSBA. In my ten years on the Burlington School Board, I've had the pleasure of attending the annual conference six times and presenting at four of them. Service on a local school board is a demanding volunteer job, and it is inspiring to spend three or four days talking with school board members from around the country about both shared and unique challenges. I'd like to thank the NSBA for the speaking opportunities (particularly the recent sexting lectures) and Bethany Kashawlic, the manager of NSBA Educational Conference Programming, for all her work in organizing the panels and presentations. I'd also like to thank Arizona educational lawyer Michelle Parker for her encouragement of this project and for inviting me out to present at the Arizona School Boards Association annual conference.

Closer to home, I would like to express my sincere appreciation to the Vermont School Boards Association and its associate director, Winton Goodrich, for their support in the development of my initial sexting lecture and my 2010 trip to the NSBA conference. I would also like to extend my thanks to Burlington School District superintendent Jeanne Collins, director of operations Terry Bailey, and director of information technology Paul Irish for providing support and assistance for an educational forum on sexting for Burlington parents and school social workers last year. All three are tireless advocates for Burlington school children, and have been a pleasure to work with during my years on the Board.

I had the pleasure of working with two outstanding members of the local law enforcement community during that presentation. My thanks to both T.J. Donovan, the state's attorney for Chittenden County, and to detective Kristian Carlson, who is a member of both the Chittenden Unit for Special Investigations and the Vermont Internet Crimes Against Children Task Force. Both T.J. and Kris are smart, capable professionals who approach their difficult jobs with care and compassion. If this book succeeds in lightening their workload in the slightest amount, I will consider it a success. Over the course of the last several years, I've had the opportunity to discuss the issues in this book at length with two other members of local law enforcement: Michael Touchette, a digital forensics examiner for the state of Vermont and another member of the VT-ICAC, and sergeant Brian Penders, a member of the Computer Crimes Unit of the Vermont State Police. My thanks to both for the interesting conversations and perceptive questions.

Over the last few weeks, I've taken the nerve-wracking step of sending my manuscript out to friends and colleagues for review and feedback. Particular thanks are due to Troy Hutchings, Mike Touchette, Mike Brunker, Haik Bedrosian, and Carol Bua Ode, for plowing through the manuscript quickly enough to provide feedback for the initial printing of the book cover. Their comments are everything I could have hoped for and more. Thanks also to the following people for agreeing to review the manuscript or galley on a more reasonable schedule for its national release in late summer: Sharon Lamb, Pegi McEvoy, Gail Dines, T.J. Donovan, Andy Rifkin, Seema Kalia, Winton Goodrich, Rep. Jason Lorber, Ann Bartow, Jane Gallucci, Peter Herley, Frank Abagnale, Lisa Shelkrot, Ray McNulty, and William Preble, Nancy Kaplan, Alison Arms, and Parry Aftab. I would also like to express my appreciation to those who have shared their encouragement and enthusiasm for the project, but are prohibited by their organizations from explicitly endorsing books: Kris Carlson, Tripp Brinkley, Jon Swartz, and David Pogue.

For independent writers like myself, the bulk of composition is done in the home, which means that family and close friends are inevitably the innocent bystanders of the creative process. I deeply appreciate the patience, support, and encouragement provided by my sons, Ben and Peter; my siblings and their spouses, Jonathan and Allison Lane, Elizabeth and Jeremy Murdock, and Katherine and Matt Van Sleet; my parents, Warren and Anne Lane; Graham Raubvogel and Emmett Werbel; Harvey and Glenda Werbel (whose guest apartment was the perfect writer's hideaway over the holidays); Keith and Penny Pillsbury; Sharon Lamb and Paul Orgel; Paul Hale and Ellen Zeman; Paul, Robin, and Michael Messer; Jeff, Marla, Amy, and Julia Liebster; Peter and Joan Strauss; Alan Matson and Dale Azaria; and Esther Margolis. A shout-out as well to my myriad friends and connections on Facebook, LinkedIn, and other social networks who have offered electronic encouragement.

For the past eight and half years, I have had the great good fortune to share an office, a house, and my life with Amy Werbel. Throughout this writing process (and three others before it), she has been an emotional support, an intellectual partner, and a bulwark against the vagaries of the writer's life. The formal dedication for each book may change, but at their core, each is dedicated with love to Amy.

Index